Energy Security for Industrial Facilities

Energy Security for Industrial Facilities

Peter A. Broussard, P.E.

PennWell Books

PENNWELL PUBLISHING COMPANY
TULSA, OKLAHOMA

Any resemblance to persons living or dead and/or institutions/companies of any sort is purely coincidental and unintentional. PennWell cannot be held responsible for any such claim.

1421 South Sheridan/P.O. Box 1260
Tulsa, OK 74101

Broussard, Peter A.
Energy security for industrial facilities / Peter A. Broussard.
p. cm.
Includes bibliographical references and index.
ISBN 0-87814-412-9
1. Industry—Energy conservation. 2. Energy policy. I. Title.
TJ163.3.B75 1994
333.79'65—dc20 93-45911
CIP

Printed in the United States of America

Contents

PREFACE

The current world energy situation makes the subject of energy security a timely topic. It is not a simple subject; it is multifaceted. It embraces efforts throughout your plant to "squeeze" the most useful production out of each kilojoule or kilowatt-hour you buy... to get "energy-lean." It requires inventory through the auditing of your energy use. It calls for prioritizing your operations. It includes protecting your facilities from sabotage, vandalism, industrial terrorism, or catastrophes of nature.

Novel energy supply strategies should be considered for options such as multifueling and waste stream energy exploitation. All of these exploratory studies result in strategic and tactical planning of your response to energy emergencies. The result is an Energy Contingency Plan, written, but always dynamic—never static.

The key to your successful energy contingency planning is involvement of all your people across the board and of respective government agencies. Involve them in auditing, recommending, strategy sessions, and gaming the plan. Invite them to participate with your corporation in influencing their legislators.

As time changes, so will your effectiveness in responding to energy security threats.

Acknowledgments and Dedication

This book was made possible only by the efforts of many selfless people who believed in me and my message enough to give selflessly of themselves. I am especially grateful to

- my beautiful wife, Mimi, for her love and unwavering, remarkable belief in my potential,
- my wonderful folks, Roy and Evelyn, for their love and tireless, thankless direction,
- my girls, Tessa and Chrysta, for their love, energy, admiration, and hopes for the future,
- my many brothers and sisters, Bill, Jack, Elly, Kathy, Tom, Angela, and Gigi who, to this day, with their own little families still teach me the art of living in love with lots of other people,

and to countless friends who helped me shape my professional career launched in no small part with some fine portrait photography by Alton Puckett,

- beginning with both Les Waguespacks (Sr. & Jr.) and "Mad Dotter" Mike Courville in the gas fields of south Louisiana,
- among the water hyacinths of Mermentau Bayou with Paul Guillory,
- along the high-voltage transmission lines of Cajunland with Leon Thompson,
- past the corridors of electric power business development aside Tom Stirling,
- through the maze of industrial Total Quality on the shoulders of Don Swift and Tim Domenget,
- to the top of the water tower of Meehan Junction sharing a ladder (and a pair of white knuckles) with Jim White,

- slipping the "surly bonds of earth" in a C-130 with Bobby Tucker,
- winding through the woodlands of Alabama at the muddy heels of Doug Robicheaux,
- scurrying around the bits & bytes of computerworld with "the Godfather," Robert Guzzino
- to the crossroads of energy security at the elbow of "the man," Col. Ed Badolato, and
- finally at the hands of Almighty God, known by many names by many peoples, who brought us all together in such a short time and made it all possible.

I extend to them all my humble, sincere thanks and love. It is to these people that I dedicate this work and to all children of future generations who must live with what we leave behind.

1

Energy Situation Synopsis

PRELUDE

On a hot August evening near Baton Rouge, Louisiana, a semi-tractor/trailer rig hauling a load of drillpipe tried to beat a west-bound freight train across an intersection and failed. The 1-1/2 mi long, six-engine train broadsided the trailer and sent dozens of joints of drillpipe flying madly through the humid air into a nearby refinery. The rigid joints of pipe easily rammed into the walls of several large fuel storage tanks, rupturing them and sparking explosions that set off a chain reaction of fires throughout the greater portion of the rest of the refinery. Other joints of the pipe were hurled into the main electric substation of the refinery. Two 9375 KVA transformers were deftly clipped of the incoming and outgoing insulators on their tops. The pipe joints also penetrated the sides of each transformer, quickly draining the fluid from them.

This was August. This was Louisiana. It was hot; it was humid. Air conditioners and refrigeration chillers were working overtime that night. The 230-kv high-voltage electric transmission lines running overhead were already reaching capacity by the time the truck lost its load of drillpipe. The fires grew quickly, feeding upon the spilled mixtures of No. 2 oil, JP-4, crude oil, gasoline, and transformer oil. It was not long after the crash that the wires failed—they melted as a result of overheating due to a combination of electric load and the fires.

These fires precluded any rapid response by fire fighters or refinery maintenance people. So the electric utility crews were excluded from the site, temporarily.

The result was that the 230-kv line was dead and no interties were possible at that substation. Nearby refineries and chemical plants were instantly without power, forcing them into emergency shutdown procedures. At many of the older plants, employees were turning valves that hadn't been turned in years, without success and sometimes with disastrous results. Shutdowns didn't go smoothly and panic further hindered operators' thinking. Tons of product were lost. Scores were injured, including the driver of the truck. Seven other men died.

The story doesn't end there. Load switching finally did take place sometime the next morning. However, it was quickly discovered that the intertied network was not sized to handle the total load. The 230-kv line that had failed was a crucial link to powering that area. Meanwhile, the utility worked heroically to restore the transmission line to service. They were finally successful...after a week and a half.

There were not enough standby generators to go around. The interties could support only a portion of the industries in the area. The utility had to work up a revolving schedule of repowering.

The shutdown in refining capacity rippled through the energy distribution industry and resulted in a general increase in fuel prices throughout much of the southern and eastern US. Some storage was drawn down enough to threaten shortages of fuel in the Northeast, which was experiencing its own heat wave.

Stockpiles were severely drawn down so that the effect of this catastrophe continued to be felt throughout the fall and into the winter.

Who could have predicted this chain of events? What could anyone (besides the truck driver) have done to prevent these things? Who could have done what, in what sequence, and within what timeframe? Who could have come out of this relatively unscathed? We'll look at some answers in this handbook.

Your Operation

The continued operation of your plant is vital to the economic health of your company and to the financial well-being of your fellow employees. An adequate energy supply is, in turn, a vital ingredient to production at your site. The energy supply to your facility can be severely curtailed or even interrupted by a host of natural disasters, man-caused disasters, and hostile activities.

Natural Disasters

Natural disasters can be defined as any emergency situation which occurs as a result of some act of nature. The most devastating include, but are not limited to, hurricanes, typhoons, tornados, blizzards, floods, severe droughts, earthquakes, volcanic eruptions, tidal waves, and tsunamis.

Hurricanes and typhoons are notorious for widespread devastation. Electrical transmission grids are extremely susceptible to this type of disaster. Hurricanes originate in the warm, moist air of the tropics and move generally northward. This phenomenon packs its forces into high winds, torrential rains, and storm surges. Windspeed within 100 mi of the eye can range from 75 to 200 mph. High winds can extend as far as 250 mi from the eye. The rainfall accompanying a hurricane averages 5 to 10 in., although 20 in. within 24 hours has been recorded. The storm surge is a massive wall of water that can reach heights of 20 ft or more. These slam into structures with horrendous impact and can literally destroy tank farms, refineries, power stations, substations, and transmission towers. Each second a hurricane is in existence, it moves 250,000 tons of water from lower elevations to the tops of its clouds. The average hurricane generates energy equal to 500,000 atomic bombs.

Although tornados are limited in the areas they affect, they remain one of the most violent phenomena known to man. Windspeeds in a funnel cloud can reach 500 mph. Because of these velocities, the air pressure within the funnel can be so low as to cause buildings to explode from the air they've trapped inside. This type of weather system is very erratic and appears suddenly, making preparation difficult.

Blizzards and extremely cold temperatures seem to target energy systems singularly. The heavy snows that accompany blizzards can cause many operations to shut down due to power interruptions or a lack of personnel able to come to work. Electric interruptions occur because the ice adds more weight to the power lines than the design limits for the conductors allow. Low temperatures can strain the capacity of gas lines to dispatch gas to residences. Gas transmission companies typically drop industrial customers in favor of residential customers. Generally speaking, severe cold temperatures are harder on equipment than severe heat. And to exacerbate the situation, cold weather severely slows down repair crews.

Earthquakes have enormous potential to devastate an energy supply network, depending on vicinity. They occur because the earth's crust is in a constant state of flux in which layers of rock slide over the liquid molten core and bump and grind against each other. At the point where maximum slippage occurs, the stress relief sends out seismic shock waves in all direction. The Alaskan quake of 1964 (8.3–8.7 Richter) released energy equivalent to 100 underground 100-megaton nuclear explosions. The effects can dramatically reduce an energy supply network to rubble within seconds. Fissures develop and large horizontal and vertical movements can result. Structures of all types, such as transmission towers, compressor stations, pipelines, tank batteries, substations, and electric generating plants can be caused to collapse. The secondary effects are often more devastating. Landslides from earthquakes in mountainous areas are unstoppable. Fire damage can also increase as a result of an earthquake. However, probably the most deadly secondary phenomenon from earthquakes are tsunamis: seismic sea waves similar in destructive potential to hurricane storm surges.

Floods can result from the accumulation of heavy rainfall or the runoff of melting snow. Ice jams have been known to block the normal flow of a river and dam it up enough to cause flooding. Also, the failure of man-made dams have caused disastrous flooding. Floodwaters carry debris which can destroy bridges, roads, and structures of all types, including power lines and powerplants. Flooding can erode soil away from and expose buried fuel pipelines, which then become vulnerable to destruction by the surging currents or floating debris. High waters can easily short out all types of electrical equipment. Floods can occur throughout the year from many types of seasonal precipitation or the failure of critical man-made structures. The duration of a flood can vary from the few hours of a flash flood to the days typical with the rise of a great river basin.

In some regions of the world volcanos are the major natural disaster. Although most active volcanos lie in the Pacific Ocean, even "inactive" volcanos need to be reckoned with, as was made evident in 1980 by the Mount St. Helens eruption in the northwestern United States. Volcanos are the result of the release of pressure which has been exerted on molten minerals lying beneath the crust of the earth. The main dangers are from lava flows, airborne clouds of volcanic ash, toxic gases, and volcanic mud flows. Although some lava flows may be slowed or diverted by strategically placed stone barricades, most efforts at controlling volcanic eruptions have not been successful.

Human Causes

There is a wide variety of human-caused disasters that can wreck havoc on energy supply streams. Some of these disasters are a result of intentional acts of industrial terrorism, vandalism, or sabotage. Others are just plain accidents due to corrosion, fatigue, changing environment, or misapplication of material. Some accidents are due to operator error, neglect, or drug use. Whatever the cause, the impact on production can be monumental.

Fires and explosions are closely related results of man-caused disasters. A fire raging into an area of flammable materials can easily result in an explosion. Conversely, an explosion can spread flame and hot debris over a large area, causing the start of major fires. Fires can affect either man-made structures or natural areas such as forests, grasslands, or prairies. Fires can be intense and localized, such as that experienced by the petrochemical industry, or cover huge tracts of land such as forests or grasslands. The huge areas cause fire-fighting resources to be spread rather thin.

Explosions are extremely violent and can cause huge amounts of damage. Especially susceptible are petrochemical storage, treatment, or manufacturing facilities.

RELIABILITY

Accidents, unfortunate chains of human errors, and severe weather can affect the energy supply to your plant. But you don't even have to look to the far-fetched to find reasons to pursue isolation measures for keeping your facility producing through it all. Just look at the reliability of the electrical power to your industrial facility. Just how good is it? How good is your fuel supply to the power plant you operate? How many fuels can your facility burn? How easily can you get these fuels? How fast? If that's not enough to worry about, consider the world energy situation. The very availability of many of the fuels you have available are tied to the attitudes, aspirations, and ideologies of people who are many miles away who don't even know (or care) if you exist.

The energy situation for industrial facilities in countries that are net importers of energy is extremely complex...and is ever-changing. At the core of this situation is a precarious dependence upon "unsure" sources of oil. You, as the manager of one of these facilities, have to provide for an ample supply of energy to your plant and that is getting to be harder to assure as time goes on.

Energy is vital to your operations. It can be found flowing throughout your installation in various forms and qualities. One thing's for certain: it makes the difference between your plant being a money-maker for your company or a lump of idle capital. Your plant is unlikely to enjoy a captive supply of energy. So you look for fuel and power, along with hundreds of other facility managers, in national and international oil marketplaces. And as time goes on, the market is running out of cheap and abundant "groceries."

So what can you, as the primary energizer of your facility, do to make sure that you have a continuous source of energy, so vital to production and profitability? It's not as entirely out of your hands as you may think. This book can be whatever you want it to be. It can be a dust-catcher, or something to justify those fancy new bookends you've just bought. Or it can be the handbook it was intended to be...the one that you use to tie all those loose ends that keep your plant from being buffeted about at the whims of a dictator who keeps getting sand in his coffee.

Operational Survivability

In this energy supply security manual, we'll look at general industrial plant survivability from a power-assurance standpoint. The ability to run your plant depends less upon your corporate affiliation than it does upon your locality and its proximity to different energy sources. Energy security is very regional. Energy security is centered on the integrity of your supply routes and the methods you use to convert your fuel into something you can use in your production.

Long Arm of the Government

The uncertainty of energy supplies are by no means exclusively a result of the actions or decisions of the cartel in the Middle East; remember that the oil industry was born in the United States and most of the technology used to develop this resource was invented here. We used to have a magnificent domestic oil industry which kept tens of thousands of our people employed at high-paying jobs. But this domestic oil industry has recently fallen victim to the inability of our representation in Washington to make some tough decisions about a long-term national energy strategy.

Our government's energy policies are often determined by politicians who are interested in short-term gain. They are continuously listening to a constituency that can't decide if it wants high oil prices and a strong domestic petroleum industry to bolster the economy or, low oil prices and *no* domestic oil industry and the recessionary trends that accompany that scenario. And energy makes such a convenient target for political exploitation: it's a capital-intensive industry with only long-term payouts for these upfront investments; the "pain" is immediate while the benefits (and evidence of wise planning) remain in the distant future. What politician would take the flak of a distasteful action just so someone in the next administration can get all the credit when the benefits finally come to light?

So, no hard, credible energy decisions are made often (as is evidenced by the arduous journey of the recently passed National Energy Strategy in October 1992). The haunting reality is that *no* decision is indeed a decision after all. So we, as those who depend upon energy to feed our families, become hapless victims of political short-sightedness. This book is an attempt to help you cope with the uncertainties.

Intracorporate Support

Increasing the energy supply security at your facility is an involved process which encompasses many steps and facets. It is an effort that needs to penetrate the rank and file from top to bottom to be successful. It is an effort which is serious enough to demand visible, unwavering support from you and broad enough to involve your people closest to the process. It is a *preventive* program, *not* a cureall to be applied after the fact. It is a detailed, sometimes tedious, dogged effort that needs to be ram-rodded by people with the right amount of "clout." It is an *ongoing* effort since nothing abroad nor locally, for that matter, is static.

If you don't want to stay in business, ignore your energy supplies.

If you're concerned about where your next kilowatt or btu is coming from, read on. We'll look at ways to anticipate the probability of a shortage of a fuel and what things you have in your power to stave off a possible shortage. We'll examine the single most potent device you can employ in this fight: methodic energy contingency planning, complete with periodic gaming and cyclic updating. If you come away from this handbook not only convinced of the need for an energy contingency plan, but determined and able to assemble one and put it in place, then this writing has fulfilled its purpose.

Me, Vulnerable?

You won't find many statistics in this book indicating how dependent most industrialized nations have become upon sources of foreign oil. You'll notice that scant mention will be made of reserves in this nation or refining capacity in that one.

These are details behind the single most glaring premise upon which interest in this book is based: that no single industry in any heavily industrialized nation in the world should be satisfied that energy, the very life-blood of its existence, will always be available in sufficient quality and quantity to keep production alive and well.

There are a host of different reasons that oil pumped from a reservoir under the sands of the Middle East will not complete the arduous journey that lays before it. It will be cradled in the belly of a single-hulled tanker meandering through such uncomfortable tight spots as the Straits of Hormuz, the Suez Canal, or the Cape of Good Hope; unloaded at a Louisiana port during the mosquito-filled night, catalytically cracked in a refinery; pumped through miles of underground pipeline to a tank farm; and hauled by an 18-wheeler across well-worn rural highways to your plant, where some individual pumps it into your storage tank. Can you feel a bit of your fuel security slipping?

The simple fact is that despite the decades that have passed since the first Arab oil embargo in 1973, a disruption or curtailment of oil from this area remains the single most potentially disruptive threat to energy consumers worldwide, be they commercial, residential, institutional, or industrial. No matter how isolated you manage to configure the energy supply for your facility, you can never isolate your plant from the effects of fluctuations in foreign and domestic oil supply. Oil is a worldwide commodity that is priced according to the ancient laws of supply and demand. Agreements exist between oil-producing companies and their ordinary customers and between producing and consuming nations such that during a crisis it is likely that part of the supplies available to the United States during a shortage will be diverted to other oil-consuming nations.

SURVIVABILITY

In this handbook we will look at industrial plant survivability from an energy supply standpoint. The ability to run your plant depends upon your locality and its proximity to different energy sources. As I said earlier, energy security is simultaneously a global and very regional consideration. It is centered on the integrity of your supply routes as well as your methods of energy conversion.

On the average in the United States, the energy situation is not real bad; on the average, that is. But "average" here is like sticking your hand in a pot of boiling water while holding ice cubes in the other—the average temperature is rather moderate, but you're not really comfortable, are you? Similarly, if your plant happens to be the only plant in the free world to have its energy supply disrupted due to an act of sabotage, industrial terrorism, or natural disaster, as far as you're concerned, despite the excellent worldwide incident rate, yours is the full 100%.

So let's get on with getting a grip on what keeps your plant going.

If you're still scratching your head wondering why you can't recall the event involving the train and the truckload of drillpipe, it's no wonder: that story doesn't make up any actual interval of history, but it could likely become a part of the future. All such scenarios that follow are fictional, and any resemblance to any events or persons living or dead is coincidental and unintentional.

2 Energy Management

Prelude

The fog was heavy at the entrance of the small polymer plant on the eastern seaboard of the Carolinas. The facility energy manager drove his rust-trimmed 1978 mini-pickup into the employee parking lot, just like he'd done time after time over the past four years. He didn't notice anything extraordinary, but the fog was really hanging around like it had nowhere else to go. What he did notice, as he walked the 50 yards to his office building, was a new blue sedan with rental car plates. That didn't disturb him, though; visitors and vendors were always coming in from out of town and calling on the plant in rental cars. What did disturb him was the crowd of people that had gathered in the lobby apparently awaiting his arrival. He thought he recognized some of the outsiders. He had no trouble noticing his boss, the plant manager and an often-seen sight, the maintenance director.

The usual half-minute was taken to introduce everyone. The plant manager then launched into a barrage of questions: How much was steam costing these days? Did we get back most of the condensate to the power house? How much of our electricity did we generate on-site? What fuel did we use the most of? Which fuel cost the most? Did he know what our power factor had been running? Were we on the best electric rate? What portion of our final product cost was energy? What programs did he have in place, or had at least started, to reduce the energy costs in our final product? Etc.

The energy manager stumbled and stammered through it as best he could from memory while he subconsciously lamented not having drank much coffee that morning. Finally, the plant manager sensed his bewilderment and filled him in as to what had precipitated the rapid-fire energy quiz and the congress of strangers convened in his office. Indications were that steam usage had skyrocketed over the past several hours, beginning at quitting time the day before, and the two gas/oil boilers were being fired to their thermal limits. Production was fairly steady, unchanged from yesterday. Yet, the boilers were still struggling to keep up with demand. The situation was serious enough for the manager to call for help from the corporate engineering office in a neighboring city about 75 miles away.

The plant manager dismissed everyone so they could prepare for a brainstorming session in half an hour to discover the sudden increase in plant steam demand. The concept of analytical troubleshooting was still a foreign one to these people, so this was more an act of desperation than calculation.

As the energy manager went back to his office, he couldn't shake the feeling that there was something he should have shared with the group but didn't. He passed the coffee pot and poured himself a cup to bring to his office. He howled in pain as the hot coffee ran down his hand from a small hole in the bottom of the cup. Not many expletives were deleted at that instant while he dabbed a napkin clumsily around his wrist. As if he didn't have enough to worry about what with steam demand running amok....

Then something clicked in his memory. He rushed back to his office and fumbled for the light switch (only then recalling that 55% of the plant's electricity was self-generated, something he wished he'd remembered only moments before). He knelt down and dragged a dusty box from under his layout table, a box full of stuff he'd salvaged from his years as an engineer in the powerhouse. He flipped through dusty yellow pages with old rusty staples clamped in the upper left-hand corners. He was not sure what he was looking for but was hopeful that his subconscious would nudge him into recognition when it appeared.

Then, there it was: "Energy Management Opportunities of the Hopewell Polymer Complex." It looked a lot older than the five years it really was. The sheets kind of stuck together as he opened it to the table of contents. It was a litany of various ways to save compressed air, water, steam, and electricity. On page 44 was described a recommendation to replace the end plates in the main condenser on the back end of the turbogenerator. During the energy audit, the investigators found that the groundwater used for cooling had a low pH and was causing pits to form near the tube welds. The welds were in danger of stress cracking due to thermal cycling of the condenser.

When he'd first read this about three years ago, he had wondered if anything had been done about it. The stress cracking had nothing to do with energy directly, but the survey team had discovered the pits while the condenser was down for a routine inspection. He remembered hunting through funded project lists and never seeing it anywhere. Because money was tight, the condenser was expensive to take out of service, and a problem had not surfaced, the project had never been approved for budget.

He made a few phone calls to the operations people on shift to read some flow rates, pressures, and temperatures. Sure enough, what had gnawed at him as a suspicion was surfacing as a reality. It seems that the pitting and stress had compromised the integrity of the tube sheet, and condenser cooling water was contaminating the condensate stream. This contaminated water was dumped to the sewer by automatic dump controls that maintained a high purity of boiler feedwater. The energy manager surmised that the cold, demineralized makeup water was requiring a great deal of steam preheating prior to hitting the intakes of the boiler feedwater pumps.

At the brainstorming session, he presented his theory, and backed it up with the data he'd gotten from the process people. The plant manager ordered the condenser taken out of service and the leaks verified. He also directed the energy manager to distribute copies of the old report and find out which of the recommendations that had been made five years ago had been put into effect.

What a way to learn a lesson! They'd paid for a study years ago, only to have it end up in a dusty old box under a layout table. Yet, this happens time and time again throughout industry. Too often the analysis of a problem becomes the final solution; it's certainly expedient and relatively cheap at the time; but, then the bill comes in later in the form of lost production.

Energy Opportunity Assessment

So what does a thorough energy opportunity assessment program look like? It contains many pieces of a grand puzzle that have to come together for any lasting benefit to accrue. One way to decrease your vulnerability to an energy supply disruption is to simply use less of it; this is cost-effective and independent of the type of energy your plant requires. Reducing energy consumption per unit of product is a vital step and should be done concurrently with the other initial energy isolation measures mentioned before. You need to get and keep your plant as "energy lean" as possible.

Waste Streams

Waste streams take on significance in plant operations, especially in securing your energy supply. As I mentioned before, one way to enhance energy security is to use less energy in the process. More accurately, you want to cut down on *primary* energy use; for most plants, that means *purchased* energy.

Waste streams from a plant contain different qualities of recoverable energy. The more energy you can recover from waste streams, the less primary energy you must buy. Some locations of these energy streams are exhaust stacks from boilers, incinerators, and internal combustion engines. Hot wastewater streams from condensers, rinsers, and washing operations contain recoverable energy. Another source to consider for low-grade thermal energy is ventilation exhaust streams from such areas as mechanical rooms, process areas, kitchens, and laboratories. Waste heat is generated by many types of equipment that may lend themselves to recovery, such as motors, generators, compressors, internal combustion engines, and gear drives. Sometimes it may even be possible to capture the heat from chemical reactions within your process.

Waste streams from your plant also serve as indicators of your process efficiency. You can use these streams as such. If you can decrease the temperature or flow of a waste stream from your facility without sacrificing product quality and quantity, you have enhanced your energy security.

Measurement

You can't control what you can't measure. So the management of your energy consumption requires measuring all of the energy streams throughout the plant, such as electricity, bottled gas, waste fuels, coal, natural gas, and fuel oil. You will also need to measure major water flows into, within, and out of the facility, such as raw water, process water, wastewater, and cooling water. These water flows are excellent indicators of anomalous energy usage. Besides measuring current flow rates, gather historical data on these flows for a more in-depth analysis.

Facility-Wide Energy Assessment

If you let it, it can really get to you, especially as you stand off from a distance and try to absorb the immense complexity of your industrial facility, with all of its interrelated processes and their numerous support systems. The nagging question as you watch the billions of btus billow out of countless stacks is: "Where do I start?"—an excellent question, too. The best way to tackle these challenges is to get help. What kind and from where? You have less to worry about than you might think because it's been done before, countless times.

Energy Balance Study

A comprehensive energy study begins with an energy balance study performed on your facility. This is an essential step because it forms the basis of subsequent analyses. Energy inputs are usually easy to obtain. These are measured electricity, raw fuels (and heating value), and average temperatures of incoming water and air. The energy leaving your facility is not as easily measured. You may need to perform a water balance to get the volume of water sewered if it is not directly measured. You'll also need to measure the temperature of sewered water. Probably the hardest quantification you'll face is of exhaust gases. Processes are constantly venting at different rates, temperatures, humidities, etc. But your heat balance depends on this so even an educated guess is better than nothing.

Figure 2–1 **Typical energy balance for a single process**

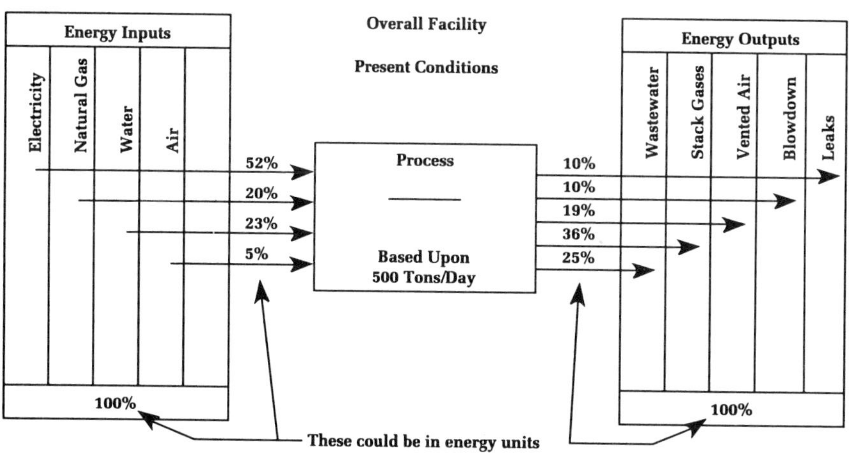

Figure 2–2 **Typical energy balance for a facility**

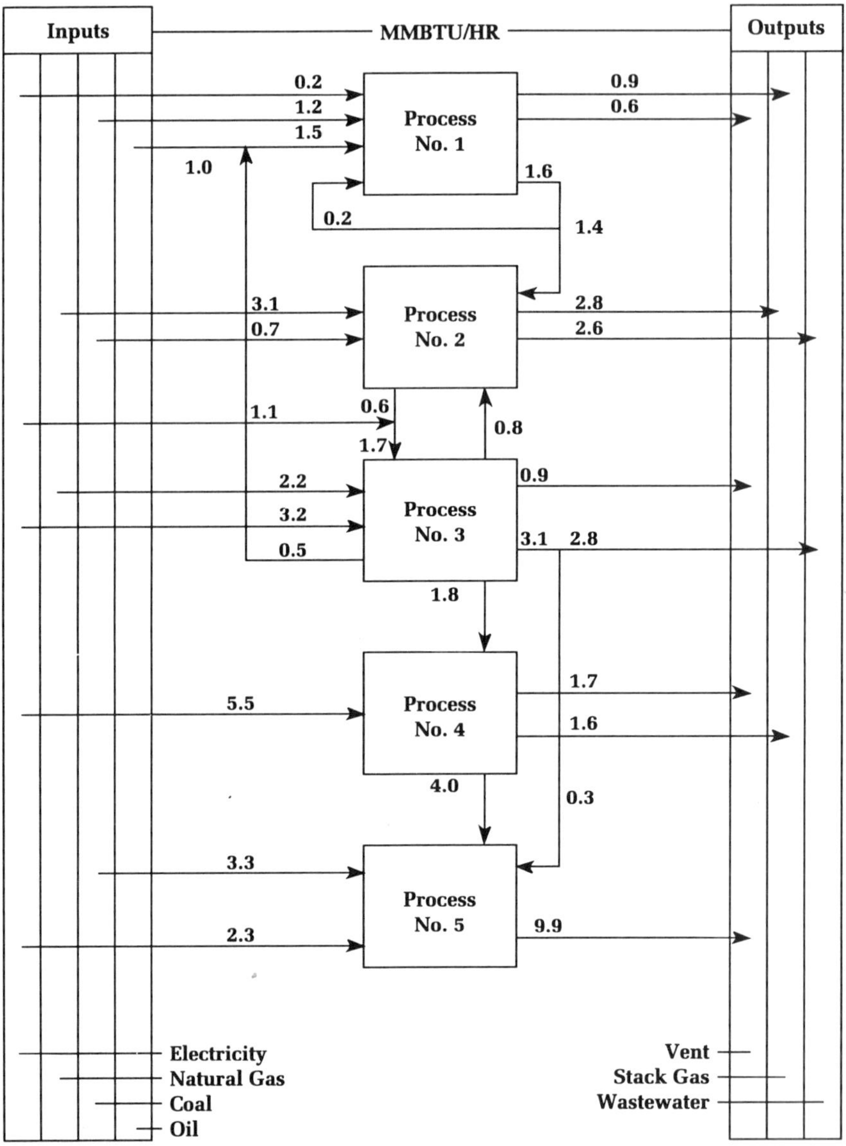

Gap Analysis

After the energy balance study is performed, you'll need to perform a gap analysis comparing how efficiently your facility uses energy to produce a ton of product with how efficient it *could* be. This step is crucial to determine the depth of the assessment.

Then, after the gaps have been identified, they'll need to be prioritized according to 1) their relative strategic importance to the plant or 2) the magnitude of energy savings possible. Strategic importance has to do with how closely a process supports production or maintains process stability. The energy-saving opportunity combines the magnitude of the gap encountered, the technological difficulty in closing the gap, and the return on your investment to install this opportunity.

Survey Organization

Basically speaking, what you'll need to do is conduct an initial survey of your plant to uncover the opportunities for energy savings, establish a methodology to get those things done, and put into place a mechanism to keep additional opportunities identified and exploited. The way that that's done is through presurvey organization, the opportunity assessment survey, and a postsurvey followup. The **presurvey organization** involves identifying and preparing all participants for what is to come during the actual survey. The **assessment survey** identifies and quantifies the opportunities available to you. The **followup** assures that the opportunities are taken advantage of and puts a continuous improvement process into place for opportunity identification.

Industrial Energy Audit

The assessment survey is sometimes referred to in the industry as an industrial energy audit and forms the launching pad for your initial penetration into energy management at your facility. It consists of an in-depth energy analysis which usually results in, among other things, an energy balance on your plant. To get to this point requires a methodical, relentless, and tedious effort into what resembles detective work. You will be looking for clues from available data to emerge with both short-term, or "tactical," and long-term, or "strategic," energy management opportunities (EMOs).

Before an opportunity identification process can be placed into operation, two key conditions should be satisfied:

1. The survey costs should be met by the facility receiving the benefit.
2. The survey program will be supported by upper management.

The first condition, while not critical, is highly desirable because it indicates that management at the local facility is serious enough about energy management that they are willing to invest some up-front money. Also, by having the audit funding coming from the local level, the facility will have "ownership" of the results of the assessment and will be more apt to act upon them than if the survey was funded at some corporate level. Finally, there is something to be said about the folks working at a plant knowing more about the applicability of suggested solutions than people who are far removed from the process in some corporate staff positions.

The second condition has a more local focus. This condition is not only critical but *pivotal* to the success of any energy management strategy, from the assessment survey to the continuous improvement process to assure program propagation. Your reading of this book indicates your support, so why mention it? Because your support as a member of upper management should be highly visible for any of this to work. Of course, you must convince the middle managers who will navigate the program through that you will support them in their decisions and expressed need for capital and expenses. But you will also need to convince the operations people of your support because they are wellsprings of ideas and suggestions.

Grassroots Support

These production people are also the folks who can make this thing work. And they *will* make it work if they know how seriously local plant management takes the management of energy. They will really rally behind you when they see how willing you are to give credit where credit is due and recognize them, *individually* as much as possible, for their efforts. Nothing works like good old fashioned stroking; it's appropriate and it's mighty inexpensive!

Person of Primary Responsibility

Other issues need to be addressed prior to embarking upon this energy opportunity survey. Among these issues is the appointment or assignment of a person of primary responsibility (PPR). This ordinarily falls to the facility energy manager. Many facilities are large enough to have an autonomous energy manager. This can be either a full-time position or an additional duty, depending upon the size of your plant or the enormity of your energy budget. Generally speaking, if your annual energy expenditures amount to more than $5 million and represent 2.5% of your final product cost, you need a full-time energy manager at your installation. If you can't justify this full-time person, then the duties of this individual will be done by a part-time person. This person could very well end up being you! If you can assign someone this part-time duty or justify a full-time energy manager, you can turn over the project of an energy audit to him or her. If it's up to you, then you've appointed yourself as part-time energy manager. What do you do as the newly appointed energy manager when faced with the initial task of performing an audit of the use of energy resources at a facility? From here on, let's address the efforts on which an energy manager needs to focus, in both accomplishing a facility energy audit and putting in place a continuing process of industrial facility energy management.

Survey Objectives

Fortunately, defining some sort of path-forward for accomplishing this survey is not hard; again, it's been done time and time before. You should begin by defining, in writing, your survey objectives. You should establish objectives around the following:

The **systems*** targeted for examination
The **processes*** targeted for examination
Typical system/process **energy consumption ranges** based upon system/process age, technology, and competition
An assemblage of **detailed energy balances** of major opportunity areas identified during the initial analysis
A **preventive maintenance/calibration program** for critical instrumentation
Additional **training** of affected personnel

*The differences between processes and systems are elaborated on later.

Guiding Principles

Besides the objectives of your audit, you should establish early on the guiding principles that will characterize your initial energy opportunity assessment survey. Some sample guiding principles could be:

1. Projects and opportunities identified must have a significant impact on facility operating costs

    - Problems will be solved vs. treating symptoms
    - Downtime will be reduced
    - Product cost will be lowered
    - Instrumentation will be immune to a lack of payback

2. Projects and opportunities will be ranked in preference by payback or other economic index
3. Results will be measurable
4. Incremental energy costs will be used for economics
5. There will result no compromise to safety, quality, nor operations
6. Additional manpower requirements will be kept to a minimum

Team Approach

The most effective process for continuing an energy management program in place involves a two-tiered team approach. One team, usually composed of people in upper plant management positions, sees energy management as but one of the many facets involved in running the facility. Remember, energy is a portion, albeit a volatile one, of the final cost of your product. Despite this "dilution" of focus, this Energy Implementation Team, as it may aptly be referred to, is the *driver* for the overall energy effort. Usually these people can facilitate the coordination of information and talent that are vital to a successful program. Another thing is that they are probably in a good position to either fund energy-saving opportunities or let you know, for instance, that there's no money available for this particular process that's outdated and is scheduled for the scrap heap, etc.

A typical makeup for this "driver" team is:

- Plant manager
- Technical manager
- Facility engineer
- Maintenance manager
- Controller
- Production manager(s)
- Energy manager

As is probably evident, you, the plant manager, are the one who needs to put this team into place; you're the one who has the authority and auspices to do so. Of course, the energy manager should be ready to advise you during this selection, but it's mostly up to you. So much for the Implementation Team.

Later, it will be good to examine more closely what this group does and especially how it relates and interacts with the lower-tiered team, the Energy Project Team. The energy manager will play a larger role in putting this team into place. These folks are the "grunts" of the business. They are empowered by the "generals" in the Implementation Team and usually get the support they need to mount a significant campaign against energy waste. The Project Team has more of a *tactical* focus, while the Implementation Team has the broader, longer-ranged *strategic* focus.

Typical makeup of the Project Team would be:

- project engineers
- operators
- production supervisors
- maintenance technicians
- instrumentation/controls technicians

For facilities that are unionized, this is an excellent opportunity for hourly/salary synergism. After all, who knows the process better than the operators who've lived with them for years?

This dual-level teaming works pretty well. The Project Team is busy making no-cost process modifications and designing projects that require capital. All the while, the Implementation Team is receiving periodic status reports from the Project Team, securing funding for necessary projects, and continuously challenging the lower-level team to identify more energy-saving opportunities.

Now, before we get any further, there's another dynamic, albeit *short-lived,* team that goes into action before the two-tiered teaming is put into place: the Opportunity Assessment Team. These people are the ones who actually "audit" the plant in the hopes of finding a way to continually drive down product costs without driving down quality or production. By now, you may be *bloated* with teams. So you may try something which has worked in the past. That is to staff the Opportunity Team such that after the audit, or opportunity assessment, *it evolves into the Project Team.* This works quite well if you've staffed the team with this end result in mind.

Besides selecting major team members, one of the things you need to do before the opportunity assessment begins is to stand back and take an overall look at your facility. Smaller plants can get by with a single Implementation Team and a single Opportunity Team. However, it could be that a plant is too big or complex for a single Opportunity Assessment Team to reasonably focus on. You may want to divide the plant into logical "parts." These can be either *process*-centered or *system*-centered, depending on how you've got your operations and maintenance set up. A majority of the time, most of the initial opportunities will be in the maintenance area, which lends itself to a logical, system-centered assessment. Usually, maintenance is divided into the types of *systems* being maintained, such as *steam, electric power, natural gas, compressed air, water, instrumentation and controls,* etc. Sometimes, however, facilities have both maintenance and operations set up around some discrete *process* or geographical composite of processes, such as the *water filtration* process or *production process* for a specific chemical, in which case it would be better to fashion teams around the focal process rather than the systems that serve it.

Major Approaches to Analysis

There are at least four major approaches to analyzing the energy consumption of a facility: 1) process analysis (examining individual processes within the plant such as a dehydration unit, or a bank of evaporation units), 2) systems analysis (examining whole plant-wide systems such as water treatment, supply, and distribution, steam and condensate return, etc.), 3) use analysis (examining misapplications of energy-containing commodities, such as using compressed air for keeping a bearing cool, instead of repairing a pump), and 4) elimination analysis (examining if parts of a system or whole processes can be eliminated without any compromise to safety, quality, or production.

During the preparations for the actual survey, i.e., the formation of the teams and targeting of the systems or processes for audit, it would be appropriate for you to determine whether or not you have the technical expertise to go through this assessment in-house. It is essential that you have on-board expertise in the specific processes and systems that you will be examining. For instance, depending upon your installation, you may need someone who has expertise in steam and power generation, catalytic cracking, Kraft pulping, textile fiber chemistry, low-grade heat recovery, industrial HVAC (Heating, Ventilating and Air Conditioning), instrumentation and controls, food processing, pharmaceutical blending, etc. Other essential experience for your team membership are in the fields of energy accounting and maintenance planning procedures.

Outside Resources

You may discover, once you've looked closely at those you have in mind for participation, that you'd be more comfortable with someone who's done this type of thing before. You need to make a determination as to whether or not you'll be using the services of some "outside" resources. In industrial facilities that are part of a larger corporate organization, it may be that these outside resources would come from a central engineering department at the corporate headquarters. Or, just as likely, these outside resources could come from one or more consulting firms. Outside resources add an objective point of view that you would be denying yourself if you confined your assessment survey to in-plant participation exclusively. This is a distinct drawback.

No matter what you choose, it is highly desirable to have on board your teams one or more people who have done this type of assessment before. Preferably, a full-time energy manager (if you've been able to justify one) would have this prior experience. This prior experience is worth looking outside of the facility for.

Assessment Plan

After the teams have been staffed, the person doing the survey coordination will need to sketch out an energy survey plan that delineates the processes and systems that are targeted for opportunity assessment. The plan should identify the specifics about "who," "when," and "how." After it has been prepared, it should be distributed by the plant manager with an explanation regarding the purpose of the survey. This endorsement by the top person at the plant is necessary to encourage participation and cooperation. Just bear in mind that this whole effort will help form part of your overall energy security planning, which could be a factor in the survivability and profitability of your plant. As such, these people "in the trenches," need your support, your input, and your involvement, especially if they're in-house staff, because many of the departments the plan will impact are focused on other things besides the saving of energy.

Data

Now let's get back to the "nuts and bolts" of this audit business. After the assessment plan has been circulated and more or less accepted, you need to get the Opportunity Team collecting preliminary energy information. They'll need to review any pertinent files and gather equipment and process design data. This may involve having to approach the vendor if the information is not at the plant site.

This design and performance data can easily serve to formulate energy use targets. The targets can then be compared to the actual performance data. At this stage, simple, one-line flow diagrams can suffice for rough energy balancing.

Among the data that should be reviewed before the actual survey are current process flow diagrams. Again, these can be simple block diagrams with one-line flows and process direction arrows. It would be very handy to have a flow diagram for each system in each process.

Another piece of relevant data that the Opportunity Team needs to have for review is the installation five-year plan as well as any other plans which may impact the survey, such as the locations of boundaries of current construction sites (which, of course, would be excluded from this survey).

Also at this juncture, the incremental costs of the value-adding utilities should be calculated, as well as the production units to be used as the common denominator for ranking and comparing opportunities. If more than one Opportunity Team is involved in the initial audit, all teams should use the same incremental costs. After all this has been done, it's time to go out and actually perform the survey.

Opportunity Assessment Survey

The actual survey will run more smoothly and will yield better results if it were performed according to some plan. This plan should include the following:

Definition of Objectives & Scope

The Opportunity Team leader should include rather global objectives here such as:

- energy efficiency
- energy conservation
- cogeneration potential
- demand-side management programs
- alternate fuels/fuel switching
- identification of additional training needs

Procedures

These procedures will set the ground rules on how data and information is to be obtained, from whom, and what resources are available at the facility. This section should include the best time and manner for involving personnel at the facility. Special assistance, tools, safety gear, and supplies necessary should be identified here as well as how to obtain them. Lockout/tagout, vessel entry, and other safety procedures are delineated here also, as well as special pass/clearance requirements. The folks conducting the survey will need to have basic facts to weigh opportunities against each other such as standard temperatures and pressures, fuel types and values, labor rates, etc. Finally, the operating hours of the facility should be spelled out.

Resources Available

This section would outline what inside and outside personnel will be available to aid in the survey. You'd want to stick an organizational chart, a facility telephone directory, and a plant layout here, too. Specify what rooms will be available for planning, reviewing, briefing/debriefing, etc. It would be good to specify what meeting room supplies will be on hand such as blackboards/whiteboards, flipcharts, markers, chalk, telephone, etc.

Schedule and Completion Timing

The main idea here is to avoid a situation where the plant doesn't suffer a compromise to its safety, production, or quality. A typical schedule might look like this:

Day 1: Kickoff meeting
- review survey plan
- introduce team members and specialties
- learn key facility personnel
- tour facility
- examine plant energy balance

Day 2–4: Conduct survey
- perform audit & analyses
- regroup periodically as necessary

Day 5: Wrap-up and exit debriefing
- present preliminary results to key plant personnel
- schedule followup/final reporting

Day X: Follow-up
- prioritize valid opportunities identified
- assign implementation teams
- determine path forward
- issue final report
- compile overall energy strategy

Survey Logistics

This is where you coordinate the coming together of resources that will be needed for the opportunity assessment survey. Airline schedules, lodging, and local transportation for out-of-town folks are laid out here. Arrangements should be made such that all nonlocal survey team members arrive at about the same time to avoid duplicating transportation requirements, especially if your plant is rather remote from a transportation hub. In-plant facilities needed for the survey are spelled out here with appropriate times and dates.

Kickoff Meeting

This meeting is one of the most important steps in a successful energy opportunity survey. It sets the tone for the entire effort. At this meeting, participants should review the goals, objectives, scope, procedures, logistics, resources, etc. that are involved in the accomplishment of the survey. All survey team members should be present, as well as all key personnel from the facility. A very important ingredient to a highly successful survey is the presence of the plant management at the meeting. This signals support for the ongoing effort that will be needed after the survey.

Probably one of the most useful products of one of these energy opportunity surveys is a list of projects that can be undertaken that will result in a savings in demand of at least one of the energy streams you examine. Using one or more of the four types of analysis mentioned above, each team should be able to produce a litany of these possible opportunities. For each opportunity listed, the team should include a cost estimate (within 30% +/–) for the implementation of the project. When combined with the projected energy savings value, a cost/benefit ratio will result. Projects can then be presented in a matrix based on their strategic significance or their cost/benefit ratio.

Final Report

What you need to remember in designing a final report format is that it should be usable to you. How many reports have you ever seen that were written fancifully and at great cost so that after its final edition it could immediately be put to work as an efficient dust-collector for shelves? Ask yourself: Who's going to use this? Will money be available to implement recommendations? Will the results need to go to some corporate people who don't have time to read reports? Should the report be all things to all people?

One of the most directly useful ways a report of this type can be pieced together is with the intent of making it an integral part of your energy planning. Part of your long-range planning will likely involve large capital projects which you'll probably not be able to afford in a single fiscal year. So these projects will need to be budgeted for in some sort of prioritized schedule over a period of years. The short-range planning will tax your manpower resources, so the same sort of scheduling of maintenance or process-adjustment projects will need to be done, except over a much shorter timespan.

Just to give you some sort of idea of what a final industrial energy audit report would look like, the following is *typical:*

- Statement of Purpose: Explains why the study was undertaken in the first place
- Executive Summary: Highlights the findings of the audit and the recommended plan of action

- Energy Balance Comparison: Graphic depictions of what the energy flow of the plant is currently, compared to what it should look like if all opportunities are implemented
- Roster of Involved Personnel: Complete with contacts info
- Survey Description: Dates, times, and a description of the operating mode of the plant during the survey
- Detailed Tabulation of Data
- Facility Energy Balance
- Prioritized List of Recommendations: Usually divided up between maintenance/expense/no-cost projects and capital projects, each list prioritized by return on investment
- Critical Energy Needs: Immediate actions required to inoculate the plant against energy shortages
- Definitions: Units of measure, values of fuels, compressed air, water, and steam at various pressures
- Checklists used by the opportunity team
- Recommended Followup Schedule
- Appendices: Logsheets, graphs of data, etc.

Survey Evaluation

Most things get better with time. To assure that this happens with energy audits in your company, it would be good to evaluate objectively how well the opportunity assessment survey went at your facility. The best folks to canvas for this would be, of course, the people involved in the survey, as well as the readers of the final report. The following questionnaire is a good launching point for you to customize your own:

Table 2–1 **Energy Survey Questionnaire**

CATEGORY	*DESCRIPTIVE*	*RATING*	*IMPORTANCE*
Focus:	The areas and processes where real opportunity for improvement exists were addressed	______	______
Reliability:	Survey performed on time and error-free	______	______
Responsiveness:	Responded to questions, concerns, suggestions of plant personnel, and incorporated suggestions for altering the process	______	______

Table 2–1 Energy Survey Questionnaire (concluded)

CATEGORY	*DESCRIPTIVE*	*RATING*	*IMPORTANCE*
Competence:	Had the required skills and knowledge to perform the survey	______	______
Courtesy:	Behaved in a polite, friendly manner and abided by the plant's norms and safety rules	______	______
Credibility:	Conclusions and recommendations are believable and trusted	______	______
Access:	Team members were easy to approach by plant personnel	______	______
Communications:	Plant personnel were kept informed; team listened to facility personnel	______	______
Service:	The team made a real effort to understand the facility and its business situation	______	______
Value:	The team identified opportunities and provided vehicles for implementation	______	______
Overall:	Performance, expectations, etc.	______	______

Implementation of Results

All this sounds nice, but no matter how many opportunities or cost-saving ideas the Project Team finds out and puts into place, if the energy-saving device or procedure or process setting is not *kept* in place, it was all for naught. To keep the savings coming into your coffers, a concept of Total Quality Management (TQM) must be embraced throughout your facility from top to bottom, including the Project Team. So, part of your job is to empower this team with the authority to set up self-directing work teams, and equip these self-directing teams with the required training, tools, control systems, monitoring systems, (especially Statistical Process Control) and excellent, flexible utility purchasing contracts. This authority will give the people who are closest to the process the ability to make the right control decisions and act on those decisions. You should, in short, through the authority you give the Project Team, *empower* the self-directing work teams. In exchange for this, you should be prepared to pay them for their additional responsibility. By the same token, they should be willing to accept a more intense degree of *accountability* for the management and relative success of the process. A suggestion would be to tie their additional remuneration to the success of the process, such as some sort of profit-sharing arrangement. You should champion this process of continuous improvement in the Project Team.

Overall Energy Strategy

Of all the things that come under the umbrella of sound energy management, compiling an overall energy strategy and putting it in writing is the most useful thing you can undertake. The most opportune time to assemble this facility roadmap is soon after the results of the initial energy audit are available. The long-range energy projects become the objects of your strategic action plans; the short-range energy projects become the objects of your tactical action plans. The tactical action plans are used to set your facility up to realize the maximum benefit from the strategic plans. And, ultimately, the strategic plans are intended to secure for your plant an energy posture that is as isolated from power disruptions as you can economically make it.

These two arenas of focus, strategic and tactical planning, make up the bulk of your overall facility energy strategy. But equally important are two other types of planning that are part of the overall strategy, the **energy contingency plan** and the **energy information plan**. The compilation of these plans is the subject of Chapter 6.

The final arrangement of this overall strategy is up to you. A suggested outline to start is as follows:

Statement of Purpose
Analysis of Situation
Key Planning Factors
Available Assets
Facility Energy Management Structure
Strategic Action Plan
Tactical Action Plan
Energy Information Plan
Energy Contingency Plan
Expected Results of Planning
Supporting Documentation

You can add to this or rearrange it somewhat. But when you look at what it is you are trying to accomplish with the overall strategy, you'll probably see that the items above are the minimum that need to be addressed.

3
Risk Analysis

Prelude

The manager of a medium-sized crumb rubber plant awakened in the wee hours of the morning on a Thursday with a disturbing thought in his mind that he just couldn't shake. Something was gnawing at him and he was having a hard time putting a finger on just what it was. The plant was running fairly well, especially compared to this time last year, thanks to the reinvestment of a large portion of the company's profits into new, state-of-the-art equipment. The market was up and his company's share had held steady, even had increased over the last twelve months. Labor relations were normal, even positive: grievances were down and labor leaders had embarked on a mini-campaign to stress taking advantage of the continuing education program of the company. The weather was good ... or was it?

As he washed the sleep out of his face before walking outside to pick up the morning paper, the water splashing over his slippers slowly awakened those concerns he had gone to sleep over the night before. This area of the country was not in the grips of a major drought plaguing the regions to the north. In fact, he knew that he'd be cutting the grass this weekend because of the almost daily showers that had been with them since spring. It was ironic that the drought to the north could impact his wet little corner of the country, but it was a stark reality. Reality had it that the electric utility powering his plant was heavily dependent upon hydropower for its capacity. Increasingly, the news was mentioning the levels in the three major reservoirs used by this utility.

But what could he do? What was his plant facing?...curtailment? The power company had not mentioned it, or had they? How much was the plant at risk for an electric power interruption? He knew he had to talk to that energy manager guy that the corporate office had recommended so highly. They had to talk today. This situation had the potential to grow into a full-blown disaster, despite the positive atmosphere that was evident throughout the plant. Now what was his name?

The energy manager had been at the plant for about three months and had not had a chance to make much of a contribution as he was busy learning his way around, and teaching them to spell his name, "Daniel Lorenzana," as well as familiarize himself with the operation. He was very surprised to receive the visit from the plant manager, Roy Lemaire. Immediately Roy got down to brass tacks. He let Danny know how uneasy he felt about the drought situation developing in the north and expressed his concern about the possibility for power curtailments by the electric utility company that was so highly dependent on hydropower for capacity.

Danny smiled a lot as Roy spoke. Roy finally asked Danny what he was so pleased about. Danny then related to him that he had worked for a company in the past that went through a situation almost exactly like the one this plant was facing. However, the problem had not been water, or the lack of it. The problem had been an extended rail strike that left the electrical power to his former facility vulnerable to a lack of adequate coal supply.

The solution Danny had come up with involved his performing a rather detailed risk analysis of the electric power supply system that fed his former plant. As one of the results of the analysis, he ended up talking with the affected utility about possible power curtailments. Based upon what he'd learned, he then assembled a written, prioritized contingency plan which not only included which portions of the plant to be shut down first in case of electric power curtailments, but also included a prioritized list of maintenance and capital projects necessary to keep the plant operational during these curtailments by the power company.

The plant manager told him of his wishes to assemble such a plan for his plant. To Roy's mounting relief, Danny told him that he was well on the way to having one put together. He'd begun a methodical risk analysis of all the external energy supply streams into the plant. Although he hadn't had a chance to talk to the gas company yet, he had already met with the power company and discovered many things about how the facility receives power and under what circumstances the plant would be weaned from its supply of electric power. He began relating some of this stuff to the plant manager, whose eyes got wider and wider. By the time Danny had told the plant manager how low on the "pecking order" the plant was, and how prone to interruptions in case of grid compromise, especially in comparison to other industries in the area, Roy couldn't keep his amazement from showing on his usually composed face. Danny smiled and nodded at his incredulity.

"Why are you smiling?" Roy asked, amazed still.

Danny said, "Oh sir, this is a negotiable thing. We merely work things out with our electrical and gas suppliers and they'll put us on select rates that isolate us from most interruptions."

"Not without a catch, they won't."

"No sir, you're right. These are higher rates, of course. What we've got to come up with is a reasonable balance among our power options."

"For instance ...."

"For instance, we'll need to weigh the costs of less interruptible gas against carrying an inventory of fuel oil which we can substitute for curtailed gas. Then we need to compare this new cost of self-generated electricity to purchased power at the more reliable rates to see what ratio of self-generation would be optimum for us."

"You've done this before?" Roy asked, with relief easing through his brow.

"Yes, but I had my own desktop computer and this risk analysis thing can take up a lot of my time if I have to do it by hand."

"Drop what you're doing, get on the horn, and let purchasing know what you need. We've got to get a handle on this whole plant power thing before it grabs us."

What ensued was a flurry of activity around putting together an energy supply contingency plan at the plant. Roy was constantly at meetings his energy manager had set up to negotiate new fuel and electric rates. Discussions on interruptibility were held with suppliers of power and fuel. Unfortunately, their plant was a little late getting into these negotiations for a priority listing on the utility interruptible schedule. The drought had spawned quite a bit of consideration by other industrial companies in the area about the possibility of a power supply curtailment. These other plants had had time to negotiate for a favorable status for power supply contingencies.

That was the first half of the plan. For the second half, Danny needed to know, and this was crucial, what internal processes could be curtailed in case electric power or fuel supplies were in some way restricted or discontinued. This decrease in production had a cost that had to be compared to the savings they expected from being on interruptible fuel and electrical rates.

The result was an energy contingency plan for that year for that industrial facility. It took about ten months to hammer out because of all of the separate interests involved. There were the fuel and power suppliers, the internal process area supervisors, the accounting department, and the internal utilities department. It was rather complex and heavily matrixed. By the time the plan was put together and distributed, the drought and its potential for harm had passed. But they would not be caught napping the next time.

Taking a look at the plan they put together, you can see that it makes sense to look outside as well as inside for potential problems with production due to energy supply. Looking inside will be handled in more detail in Chapter 5, Multifueling.

To plan for a possible energy supply disruption, you should know how likely this disruption is to see if it's worth investing time and maybe even capital in planning for its occurrence. Redundancy, duplication, and diversification in energy supply come with additional costs which you must weigh against the costs of shutting production down. Most of the time, production shutdown is so costly that measures to stay on line can be easily justified economically. So, how probable is a loss of a particular energy source, either because it is physically unavailable or is priced totally out of reason?

Risk Analysis

One way you can gain some insight into this probability is to perform a "risk analysis" upon each of the external energy inputs into your plant. Rather than trust in someone's (or your own) subjective opinion about the likelihood of an unwanted event, it is far more objective, accurate, and reliable to submit an existing, operational system to scrutiny by this risk analysis process. Certainly, your access to necessary information on utilities outside your plant will be limited, so your analysis of risk will not be as rigorous as that which you can perform on your internal components of energy demand. But remember that you are just looking for some sort of *relative* energy supply vulnerability. Risk analysis should yield a priority list of critical energy supply components which can give you direction as to where best to focus your attention (and capital) for planning.

There are several techniques that fall under the umbrella of risk analysis. Each technique has proven strengths and weaknesses and is more applicable for some processes than others. Although there are some which are useful for analyzing systems which are under design or newly constructed, let's consider only those methods applicable to mature systems already in place and operating.

Approaches and Techniques

One of the most thorough treatments of risk analysis methods and appropriate applicabilities is published by the American Institute of Chemical Engineers in their *Guidelines for Hazard Evaluation Procedures.* No less than twelve separate methods of analyzing risk are presented and explained. These are the 1) Safety Review, 2) Checklist Analysis, 3) Relative Ranking, 4) Preliminary Hazard Analysis, 5) What-If Analysis, 6) What-If/Checklist Analysis, 7) Hazard & Operability (HAZOP) Analysis, 8) Failure Modes & Effects Analysis (FMEA), 9) Fault Tree Analysis (FTA), 10) Event Tree Analysis (ETA), 11) Cause Consequence Analysis (CCA), and 12) Human Reliability Analysis (HRA).

The first six methods listed above lend themselves more readily to analyzing systems which are being designed or are newly constructed. The results would not be detailed enough for you to analyze fully the reliability of your energy supply system. The HAZOP and FMEA are useful when considering the effect of a single component failure upon the operation of the whole system. This leaves you standing there if you haven't a clue as to which of the many components in the system will probably fail first. The FTA sort of "backs into" the analysis. It considers a single, undesirable event (termed the "top event") such as an interruption of electric power from the grid, and examines what possible event or chain of events could have caused it. The ETA more or less proceeds in the opposite direction and begins with an "initiating event" and takes the process through to the potential of an accident. It is sometimes used to target the top event for an FTA. The CCA is most useful as a communication tool and is a blend of the FTA and the ETA. Finally, the HRA focuses upon the potential for human error and its causes.

The method you finally choose to perform the risk analysis on the system you've targeted depends upon 1) the results you want, 2) the data you have on-hand, and 3) the stage of development of the system being analyzed. A fine methodology for selecting an appropriate method is set forth in flowchart form in the book *Guidelines for Hazard Evaluation Techniques* that was mentioned before.

Fault Tree Analysis

Because it lends itself to the analysis of mature systems and focuses upon a basic undesirable event, the **Fault Tree Analysis** is probably the most appropriate method to use in analyzing the risk of a potential power or fuel supply interruption to your facility. This method has been developed by a host of professional sources, particularly the U.S. Department of Energy for examining the risks of nuclear power plants. The DOE booklet, *Fault Tree Handbook*, is worth obtaining.

A thorough treatment of Fault Tree Analysis would be redundant here because it has been more than adequately covered in the references made above. It's probably not necessary for you to dive into the depths of FTA technique and theory to gain some insight as to its usefulness as a tool to look at your system. However, some basics should be covered at the outset, such as symbols used and the logic behind the symbols.

Since things seem clearer to us when presented visually, it really pays off to use graphics to express complex situations. The Fault Tree Analysis does just that. The FTA illustrates combinations and chains of failures that result in a single, undesirable failure which we're interested in. As mentioned before, this failure is called a "top event." For the FTA, this is where you start. You look at that single failure that could cause you misery, such as not having coal for your boiler or an interruption of electrical power to the bus serving your plant, and examine all of the *immediate* failures that could result in that single top event.

Once you've isolated those immediate events, you, each in turn, examine all of the immediate failures that could have caused them. And so on. You repeat this procedure until you come to a point where further examination is either 1) impossible or 2) unnecessary. The first condition could result from not having any more information at that point. For the second condition to be possible, the event you've arrived at needs to have been quantified sufficiently such that you don't need to analyze it further.

Minimal Cut Sets

And all of this is done graphically with standardized symbols and methodologies. The resulting flowchart resembles an upside-down tree (hence, the term "Fault Tree") with the base (at the top) being the top event. This diagram, with its "and" gates, "or" gates, basic events, etc. is then "solved." That is, it is used to put together a list of the combination of failures (or single failure) that can ultimately result in the top event. Each item on this list is called a "minimal cut set." This list of minimal cut sets is the final product of the fault tree analysis. Your determination of the likelihood of the top event occurring is based upon the chances of *any one* of the minimal cut sets happening.

If a minimal cut set contains only one component, it is a cut set of the first order. A minimal cut set with two components that both must fail in order for the top event to occur is a second order cut set, and so on. Your focus for minimizing the occurrence of the top event should be centered upon minimal cut sets of the first through third order. Beyond that, the law of diminishing returns begins to dominate your efforts.

Fault Tree Solution

Solving the fault tree, that is, coming up with a list of minimal cut sets, can be done either manually or using a computer program. Because most of the systems you'll be dealing with will probably result in rather large fault trees and because you'll probably be analyzing more than one system and because the manual solution is so time-consuming, you should look into purchasing some fault tree software if you are serious about risk analysis of your energy systems, internal and external. If you wish to gain the ability to manually solve fault trees, again, either the DOE *Fault Tree Handbook* or the AICE *Guidelines* ... are excellent.

Fault Tree Minimums

Whether you choose to use software or solve the tree manually, it is up to you to define 1) the top event, 2) the detail (level of resolution) of the analysis, 3) the boundaries of the system being analyzed, and 4) any assumptions or disallowed events (such as collisions with kryptonitic asteroids!) that you wish to use to help focus your effort.

One of the best ways to gain insight into this particular analysis technique is to consider an example. Suppose that you have a choice between using coal and natural gas as a secondary fuel to your oil-fired furnaces. You'd like to perform an examination of the supply networks that you'd have to depend upon to get either of the fuels to your site. The steps are as follows:

1. Assemble layout diagrams of both networks.
2. Perform a system analysis on each network.
3. Construct a fault tree based upon your analyses of the configurations.
4. Perform a permutation analysis on each fault tree.
5. List the minimal cut sets that result.
6. Perform a frequency analysis on each of the cut sets to determine the number of critical components in each network.
7. Analyze the potential threats that can befall each critical component in each network.

8. Make your decisions based upon:
   a. the likelihood of a failure of *any* critical component in a network or
   b. how effectively *you or your company* can reduce the vulnerability of each critical component on the *external* network.

For instance, a coal supply system might look like this (Fig. 3–1):

Figure 3–1 **Coal Supply Network**

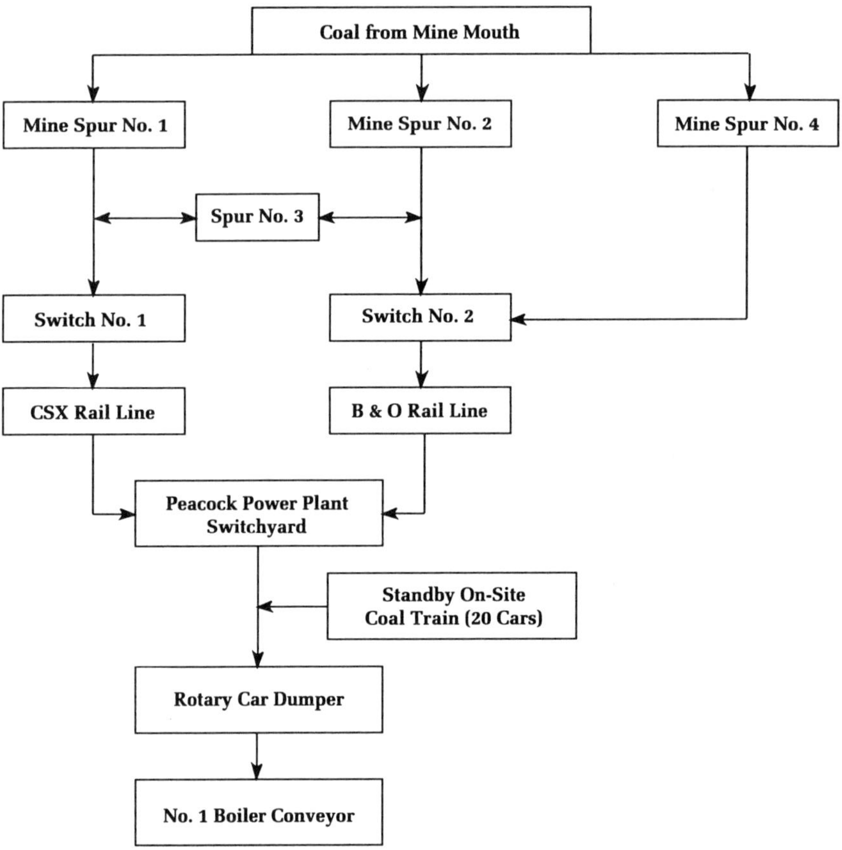

The resulting fault tree would look like this (Fig. 3–2):

Figure 3–2 **Fault Tree for Coal Supply**

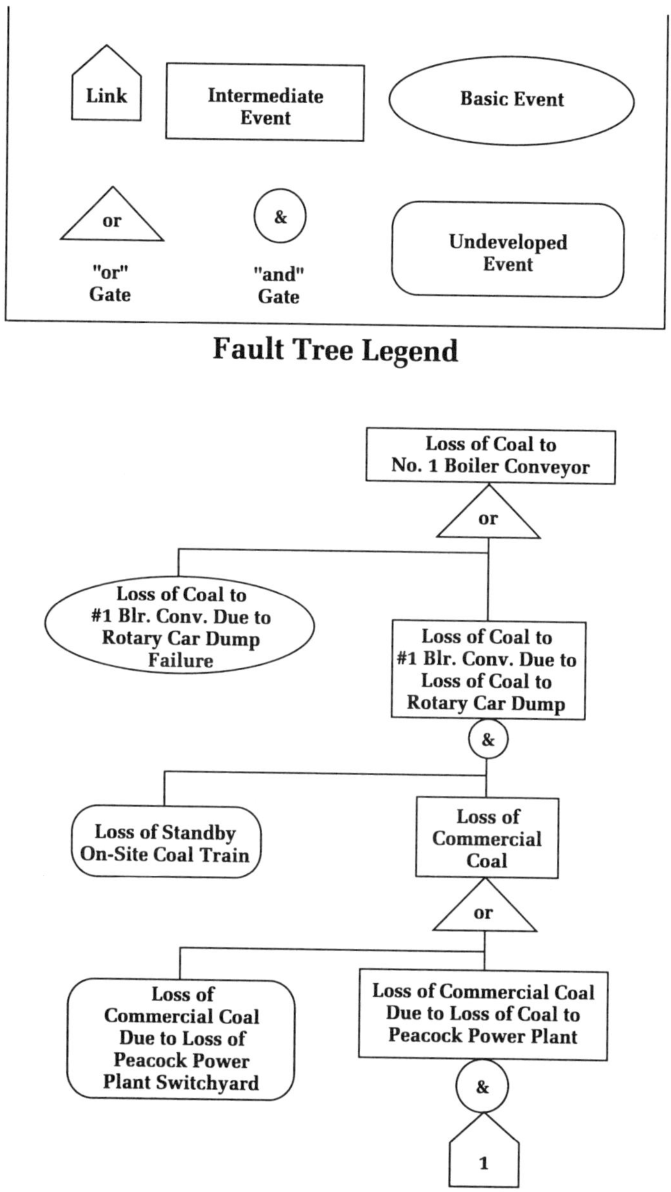

Figure 3–2 Fault Tree for Coal Supply (continued)

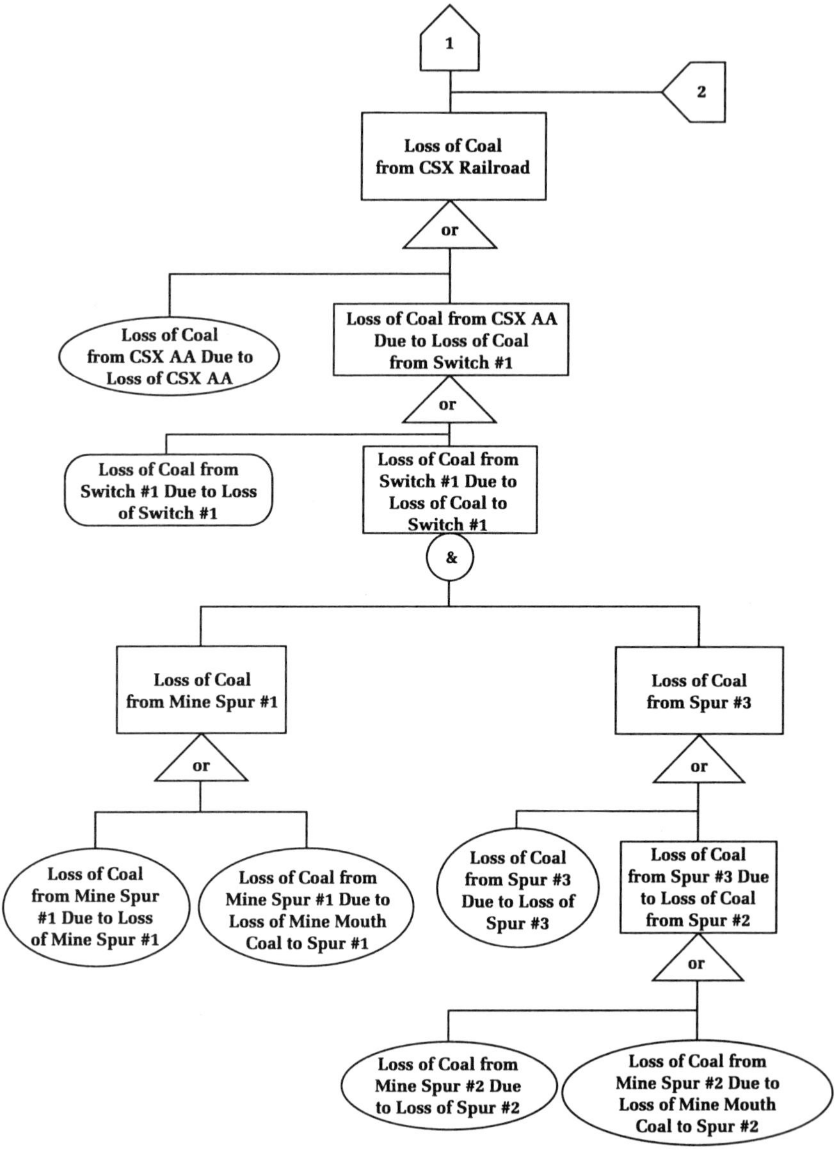

Figure 3–2 Fault Tree for Coal Supply (continued)

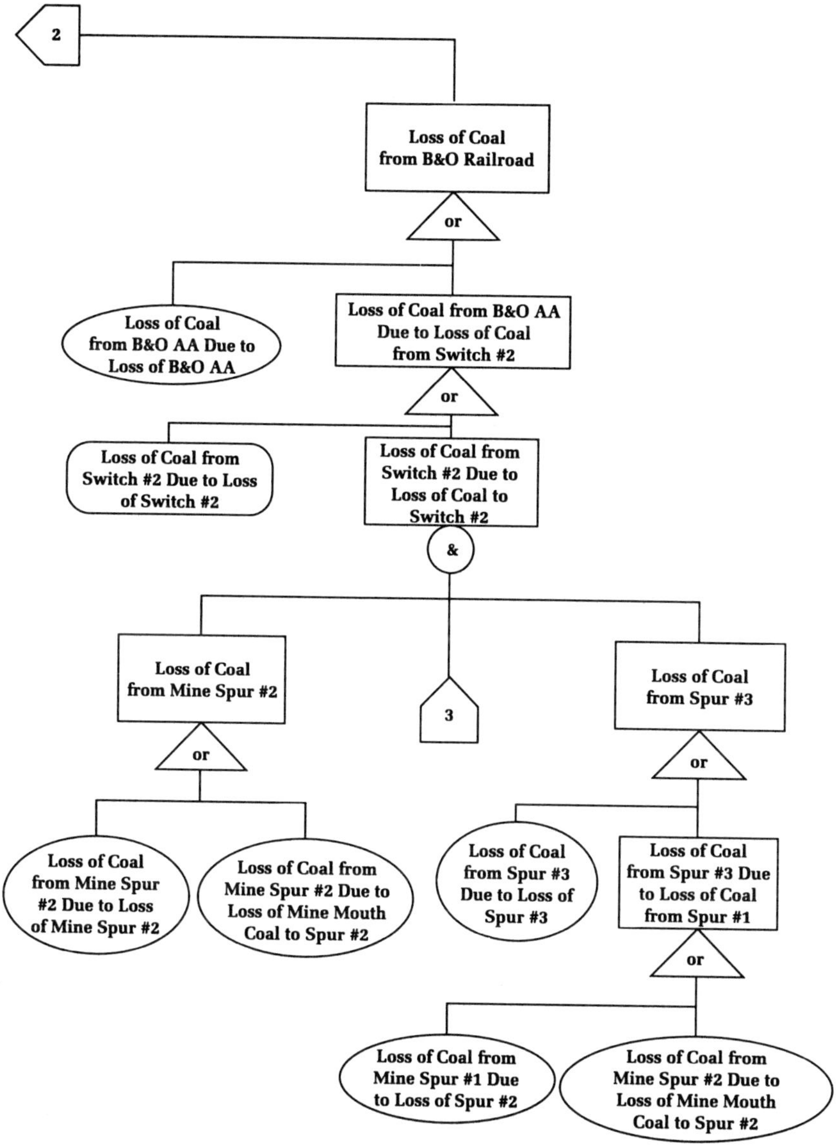

Figure 3–2 **Fault Tree for Coal Supply (concluded)**

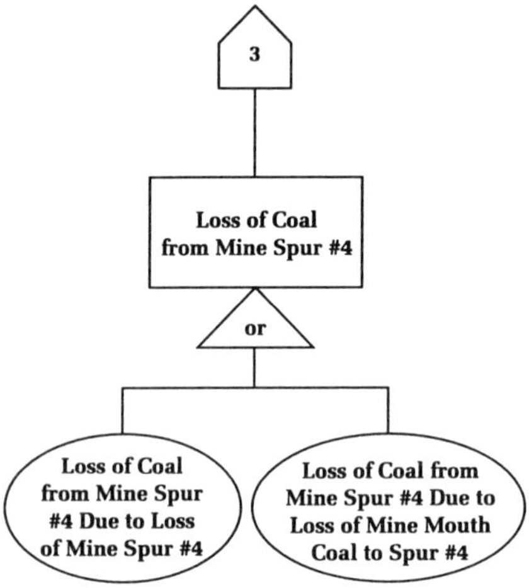

Table 3–1 **Minimal Cut Sets**

The resulting minimal cut sets for the fault tree diagrammed above are as follows:

Rotary Car Dump				
On-Site Train	Mine Mouth			
On-Site Train	Peacock Switchyard			
On-Site Train	CSX RR	B&O RR		
On-Site Train	CSX RR	Switch #2		
On-Site Train	B&O RR	Switch #1		
On-Site Train	Switch #1	Switch #2		
On-Site Train	Spur #1	Spur #2	B&O RR	
On-Site Train	Spur #1	Spur #2	Spur #4	
On-Site Train	Spur #1	Spur #2	Switch #2	
On-Site Train	Spur #1	Spur #3	B&O RR	
On-Site Train	Spur #1	Spur #3	Switch #2	
On-Site Train	Spur #2	Spur #3	CSX RR	Spur #4
On-Site Train	Spur #2	Spur #3	Spur #4	Switch #1

These sets are combinations of failed system components that bring about the loss of coal to the No. 1 Boiler Conveyor. From an examination of this listing, you can pick out the "critical components," i.e. those from the first and second order cut sets. These components in our hypothetical coal network would be the Rotary Car Dump, the On-Site Train, the Mine Mouth itself, and the Peacock

Power Plant Switchyard. These are the components which you should examine more closely for vulnerability. If they are rather improbable of interruption, then the coal supply can be considered relatively secure as a multisource fuel.

Ranking Critical Components

If you are interested in ranking the components in accordance with how critical they are, the standard procedure in doing so would be to rank the critical components first. Then you would rank the remaining components not originally selected as "critical" in accordance with how often they appear in the third and greater ordered cut sets. But in the case of multifueling, since you are comparing the relative risks among external fuel supplies, it suffices to examine the vulnerabilities and number of critical components in each fuel supply network being considered.

For instance, suppose the natural gas supply system looked like this (Fig. 3–3):

Figure 3–3 **Natural Gas Supply Network**

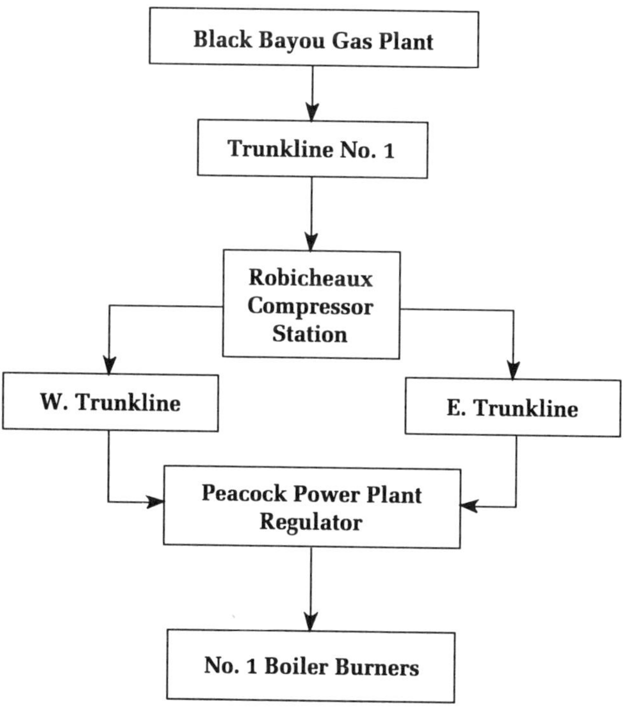

The resultant fault tree would look like this (Fig. 3–4):

Figure 3–4 **Fault Tree for Natural Gas**

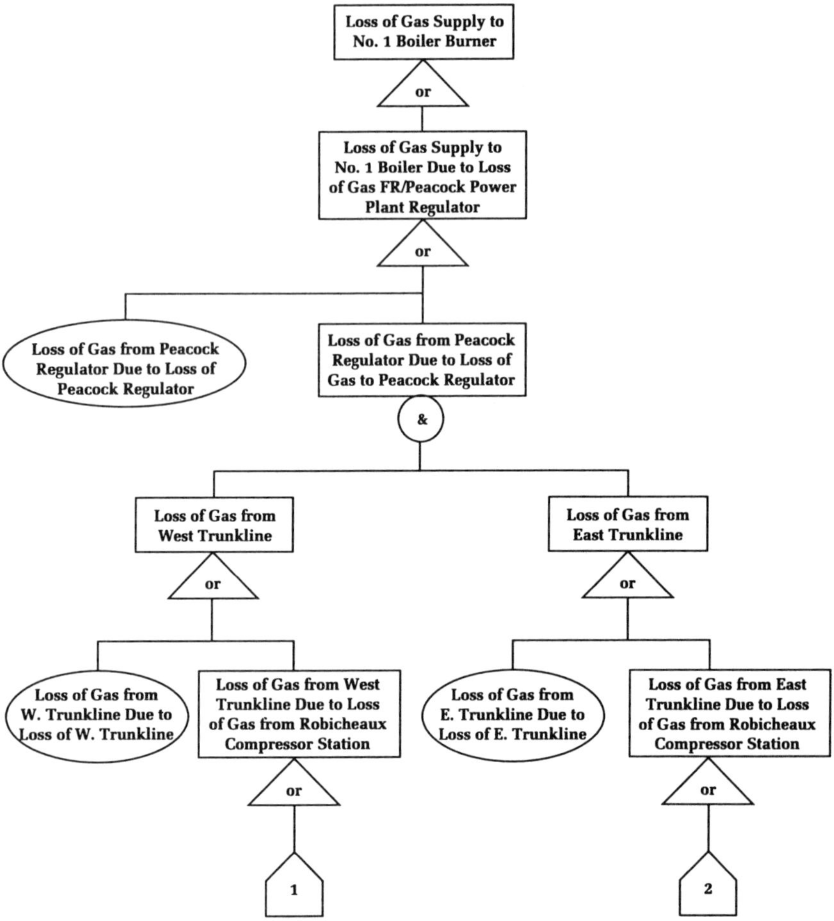

Figure 3–4 **Fault Tree for Natural Gas (continued)**

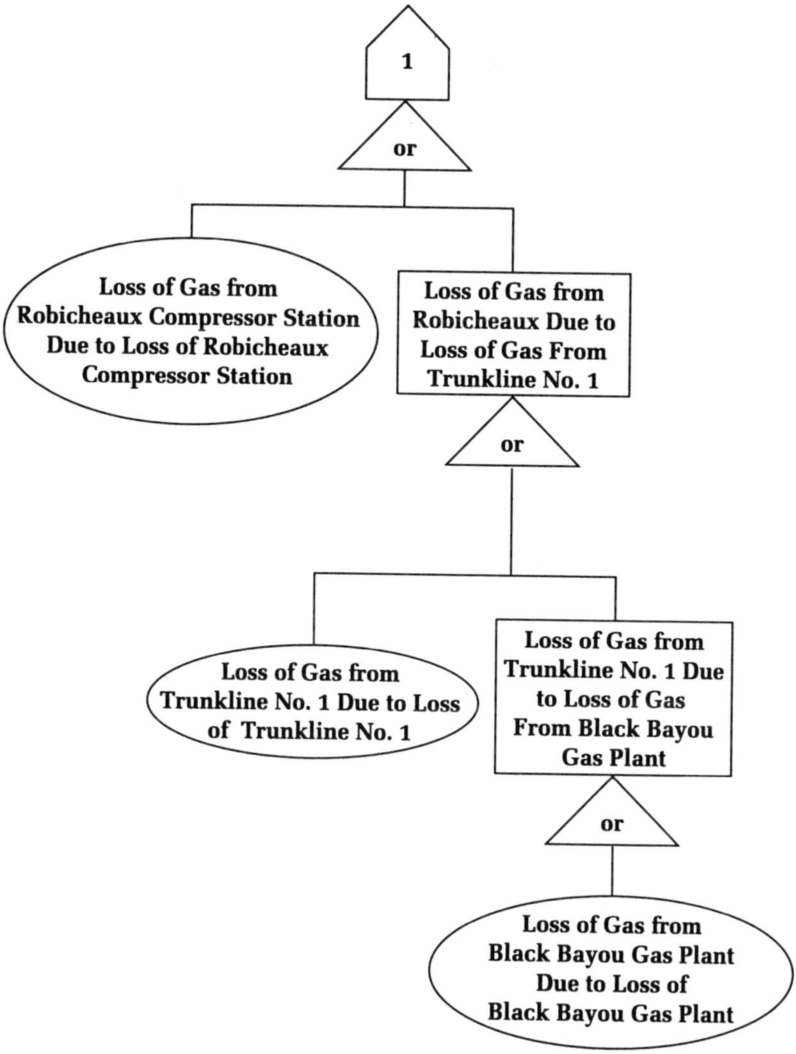

Figure 3–4 **Fault Tree for Natural Gas (concluded)**

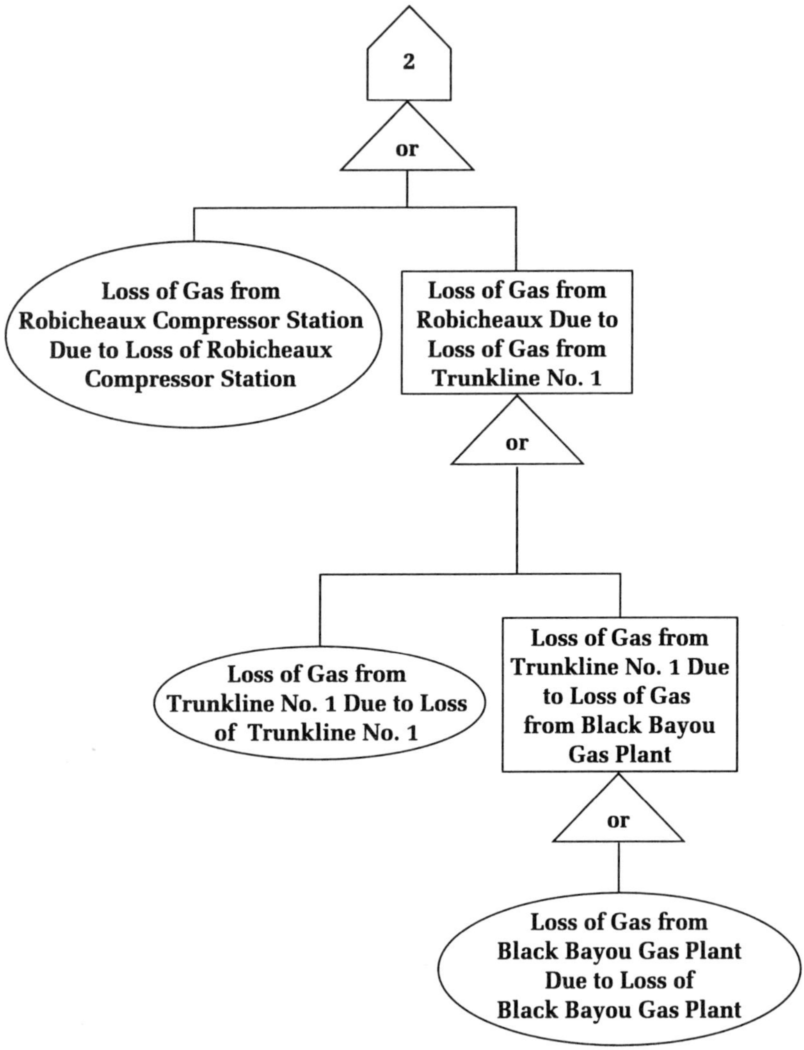

***Table* 3–2 Minimal Cuts Sets for Gas Supply System Example**

An analysis of the minimal cuts sets reveals that there are only five:

Peacock Regulator	
Robicheaux Compressor	
Trunkline No. 1	
Black Bayou Gas Plant	
West Trunkline	East Trunkline

However, four of the five are single-order and the remaining one is a second-order set. This could lead you to believe that this is a rather vulnerable system compared to the coal network, which has only one single-order and two second-order minimal cut sets. It's true that the coal system has more redundancy in it than the gas system. But what gives you a true idea of reliability is an examination of the vulnerability of the first- and second-order critical components. In these cases, the vulnerability of the *Rotary Car Dumper* is compared to the *Peacock Regulator,* the *Robicheaux Station, Trunkline No. 1,* and the *Black Bayou Plant.* The failure rate of the Rotary Car Dumper is weighed against the highest failure rate among the Peacock Regulator, the Robicheaux Station, Trunkline No. 1, and Black Bayou Plant. If that's a tie, or if either is insignificant, then the highest failure rate between the On-Site Train and the Mine Mouth is weighed against the highest failure rate between the West Trunkline and the East Trunkline.

Even though sheer numbers of critical components tends to work against the gas system, the mechanical nature of the coal components increases the vulnerability of that system. You'll probably find that the failure rate of a rotary car dumper soars in comparison to a pressure regulator. But the complexity of the compressor station contributes to its failure rate. This one would be too close to call from a "seat-of-the-pants" judgment; some actual failure data from over the past couple of years would go far to aid in your comparison.

An added benefit to performing this sort of risk analysis is that it functions as a sales tool which you can use to justify necessary actions intended to preclude occurrence of the top event. It can be used, not only to reveal to *you* where your vulnerabilities are, but to show other people on your staff as well as the people you are responsible to where those bottlenecks to secure production are. It is a rather credible analysis because it is objective and as thorough as you wish to make it.

Consultant or Software or Both

Because the use of FTA can become, for a plant on the cutting edge of its industry, rather pervasive, and because even a simple system can yield a large fault tree, you should really investigate the purchase or lease of computer software that would allow you to do a multitude of these analyses. The alternative would be to engage the services of your security consultant for these in-depth analyses. If his or her range of service is adequately broad, this type of analysis should be routine. Also, a security consultant is more likely to have computerized fault tree capabilities.

Other Uses for Fault Tree Analysis

Throughout this book there will be mentioned the need for prioritization or ranking of some sort. Almost without fail, you will see the use of fault tree analysis as the fundamental tool to assemble your priority list. The major reason for this is that FTA does such a wonderful job of imparting objectivity to this otherwise potentially subjective task. In fact, once you've done a few and gotten the hang of it, even more applications will become obvious to you.

4

Physical Security

PRELUDE

The March/April 1988 bombings by the Islamic Jihad of the 450,000 b/d oil refinery at Ras Tanura, Saudi Arabia, followed by the attack on the nearby 250,000 b/d refinery at Al Jubayl, would not ordinarily have had much impact on the folks in the little south Louisiana swamp town of Hackberry. But the terrorist attacks by the Iranian-backed Hizballah (Party of God) became the major topic of conversion amongst the buzzing of the mosquitoes before July of that same year.

Now, these mainly rural-turned-oil folks didn't possess an in-depth knowledge of the intricacies of international terrorism, nor even a cursory awareness of the multitude of different terrorist factions constantly at work throughout the world today. What they knew a lot about was what had happened recently in those old oil fields all around the place they called home. And it seems that not a small part of the bizarre incidents were directly attributed to terrorist influence.

What happened was that a newly hired man from Texas was fired after working through the heat of a South Louisiana summer as part of a production crew. Brian Puckett was not a sophisticated individual, but he was somewhat skilled. The reasons he was asked to leave are not clear, nor are they pertinent to the ensuing events. But what is important is that he apparently nursed a slow rage for some time after the fact.

Coincidentally, it was a little after Puckett was fired that the refinery bombings occurred in Saudi Arabia. The stories hit the news and seemed to ignite the Texan. The news stories and photographs describing the incidents probably went a long way in piquing his interest. And it appears that not long thereafter he began busying himself with the details of his revenge against the oil company that let him go. He saw in the Saudi Arabian refinery attacks the destructive potential locked up within the miles of pipelines there ... and in the old familiar oil fields around his home.

Hackberry Oil Field has been producing for almost a half century. The field is dotted with beam pumping units, accessible mainly by crew/service boat: this is Louisiana marshland. It was perfect for someone to scoot around without being noticed and without leaving tracks to follow.

What Puckett did in the course of a single night set off a chain reaction that rippled through the marshes of south Louisiana all the way up to the top stories of the Kayne Building in Houston, the oil company's headquarters. All it cost him was about $248, including a new ax, a cutting torch, acetylene bottle, an oxygen bottle, and some gasoline. (He borrowed his brother's boat and motor.)

After dusk on Saturday, May 21, 1988, Brian Puckett slipped into the dark, humid, mosquito-filled marsh and stealthily made his way to the first of twelve beam pumping units he would see before morning. He didn't need much light; his battery lantern was all he used.

Beam pumping units consist of an electric motor or natural gas-powered engine turning a large gear box by means of a set of V-belts. The gear box sets in motion a "ballet" of moving parts, including the crank counterbalances, the connecting rod or pitman, and the walking beam which rocks back and forth on a steel support called a Sampson post. At the head of the walking beam is the horse head, around which are wrapped the pair of cables that pull on the polish rod. All of this motion is necessary to move the subsurface pump downhole in the oil-bearing formation through continuous up-and-down cycles. Counterbalances are necessary because the downhole pump and sucker rod assembly are considerably heavy.

Thousands of this type of unit are planted across the world. In East Texas, near Spindletop, you can watch the sun set on a herd of these "rocking horseheads" as they continue their nodding up and down, up and down....

Puckett stopped at the first unit. He swung the ax mightily, eventually hacking through the electric conduit to the motor. The electric fireball that burst into flames around the head of the ax caught him off guard and he fell in his clumsy attempt to run. He then remembered to put on some safety glasses. The motor stopped and he hacked through the screened belt guard until it gave way. He managed to cut through the four V-belts that turned the gearbox. Then he rested. The hum of another beam pumping unit nearby mixed with the lapping of the water against the platform piles soothed him and gently chattered approval of his rising campaign against the oil giant. He listened for the sound of a boat: nothing.

A mosquito whined into his ear canal and he slapped himself sharply. It was enough to rouse him into turning his attention to the next matters at hand. He lit the cutting torch and within less than a minute had cut the sucker rod clean through. The entire rod string and downhole pump plummeted to the very bottom of the wellhole.

He beamed at how absolutely effortlessly he had wrought immense harm to the company. He broadened his smile and imagined himself a knight, standing over a sleeping dragon after removing its teeth. He breathed in a renewed sense of purpose and determination. He piled his tools into the boat and darted to the next pump. He repeated his destruction with budding methodology; he finished quickly. As he did this again and again, he took less and less time to complete his task.

There was absolutely no one else aware of his activities. What eventually overcame him was his own fatigue. He became more satisfied with his vengeance as his arms grew stiffer and the blisters on his hands begged him to stop.

At last, after completely immobilizing twelve pumping units in the span of five hours, he quietly piloted the bass boat away from the oil field, pleased that he'd wrecked sufficient compensatory havoc upon a company that had dealt him what he considered an equal blow.

With each pumping unit disabled, crude oil production dropped at the field. The reduction came in stages, though, and went on unnoticed by the production office at the field. The first sign of trouble was a horrendous fireball that lit up the marsh like a sunrise.

Bryan Puckett had been disappointed that his activities at the beam pumping units didn't have the brazen flair of the terrorist attacks in Saudi Arabia which had inspired him so. He longed for a suitable sign to underscore his anger and his newfound superiority over a company that had demonstrated his inferiority.

He wanted to go beyond annoyance, beyond fiscal damage...he wanted terror. He began by taping a scrap of paper on each well he sabotaged with a Star of David and the word "Hizbala" scrawled upon it, completely unaware of the dichotomy of those two symbols and the nonsense he flaunted.

After he had finished with the pumping units, he made his way to the tank farm where weeks of production were stored in a battery of six 25-year-old bolted steel tanks. He cut off his outboard motor and paddled to the shore. He selected the largest tank; it would yield the most damage and was the easiest to hide behind.

He quietly hauled the cutting torch and bottles to the base of the 100,000-bbl tank. He opened the valves on both bottles all the way. After he lit the torch, he set the flame all the way past a "heat" flame to a "cut" flame. He tied off the torch trigger with an old piece of wire from the bottom of the boat, laid it down next to the crude oil tank with the flame pointed at the metal skin, and slipped away. He paddled quietly but determinedly until he was quite a distance away. Then he started the outboard. He knew that it would take some time before the torch finally cut through the tank.

Although he was glad to be back at his truck hitching up the boat so that he could make a quick getaway, he was a bit disappointed that nothing had yet happened at the tank farm, even though he had no idea what to expect.

The torch eventually cut through the tank. The ensuing crude oil leak was a trickle that caught fire immediately upon exiting the tank. But it was a small enough fire to go unnoticed for awhile. Before long, the bottles lay in a pool of fire. The hoses burned in two and let high-pressure acetylene and oxygen jets mix with a fury. The steel plate got so hot that it failed at the plate seams, which allowed more oil to spill out.

The night hid the black clouds of smoke. And the glow of the intense acetylene fire, as brilliant as it was, also went unnoticed due to the curtain of heavy smoke and its location behind the large tank. All the while, the bottles lay engulfed in flames until the heat accelerated the buildup of pressure. The bottles failed. The explosion of oxygen and acetylene ripped open the side of the crude oil tank and spewed flaming crude over a 100-ft radius. Heavy chunks of steel from the gas bottles shot into two nearby metering stations for high-pressure natural gas. The escaping gas burst into brilliant jets of yellow-blue fire.

The crew on duty that night were caught off-guard and worked in a state of semi-shock. The roaring of the high-pressure gas drowned out the yell that Puckett let out in his glee and total amazement. Before too long, he collected himself and drove quickly away from the scene of his handiwork.

The damage was almost indescribable. Because the oil field was located in a remote area, the fire-fighting capacity needed to extinguish the flames was not on the scene before the fire spread to the adjacent tanks. In all, three 85,000-bbl and one 100,000-bbl tanks succumbed to the initial fire; almost 2 million gal of crude oil were lost.

The "Hizbala" slips of paper were too shallow to throw off the authorities, but the locals lapped it up. They became convinced that some OPEC country had struck in the middle of the night in an effort to slow down U.S. domestic production. Within two weeks, the little public library in Cameron, Louisiana was packed with folks hungry for knowledge about the Middle East and its inhabitants.

Meanwhile, putting the production site back together took many months and much capital. The losses in energy went beyond the 2 million or so gal of crude oil and the 500,000 MCF of natural gas that burned. The wells took weeks to repair. The downhole pumps proved stubborn to fish out. But even after they were repaired, they were shut in along with most of the other wells. Crude oil from the remaining storage tanks had to be emptied onto barges to make room for accepting field production from wells that had been shut in for too long. Even with the two tanks empty, field production had to be curtailed.

Facility Security

As with many major energy incidents, the repercussions of the curtailed oil production from the Hackberry field reverberated throughout industries along the Gulf Coast. Refineries, lube plants, chemical plants, and, ultimately, fuel distributors felt pressure from the production slowdown.

The physical security aspect of energy contingency planning is a broad subject unto itself. Although we are herein concerned with a narrow scope of security, i.e., energy supply, making your energy supply physically secure goes hand-in-hand with making your plant physically secure. The majority of security measures to be considered will center on protection of facilities against actions perpetrated by man. But the precise degree of protection depends on what type of person is taking physical action against your plant and where this perpetrator will strike. In other words, know your enemy and your facility before you plan your defense.

Some industries are directed by the government as to how much security to use, such as nuclear electric generating plants, ammunition plants, and some military installations. But most facility managers enjoy great leeway as to how much security to employ.

Some semblance of security already exists at most facilities. This is mainly to protect companies from theft, vandalism, loss of propriety, and to insure personnel safety. What usually happens at a plant is a gradual relaxation of security measures to a level which allows the guard to stay in his shack and wave at people entering and leaving the plant. This is unfortunate because nominal gate security does not come close to protecting your facility from a rather destructive phenomenon we see occasionally, particularly in developing nations. That phenomenon is industrial terrorism.

Some estimates claim that industry spends, on the average, $100 million yearly from direct damages from terrorist activities; the indirect financial costs of these attacks are practically immeasurable.

Virtually nothing is safe from a violent attack from someone dedicated to its destruction. Any energy facility or fuel line could be targeted by terrorists for any number of political, personal, or pseudo-religious reasons. Compounding the problem, most of these facilities are not guarded sufficiently to preclude isolation from terrorist attack. In some cases, a catastrophe can be caused by the simple turning of a valve or the throwing of an electrical switch. Primary targets of terrorists are sites which have the greatest potential to affect a wide portion of the population. Municipal and industrial facilities that handle, store, process, and consume chemicals are highly targetable, as are industrial and municipal facilities that provide drinking water or food.

The security methods you employ will depend upon what energy stream you are protecting, its accessibility, the type of person likely to attack it, and how the attack will probably occur. You will probably conclude that a wide variety of potential saboteurs and targets exists. As stated before, it would be impractical to guard all energy streams, at all times, against all adversaries. Even if you were to beef up your effort and launch an all-out bid for physical security of the energy choke points at your plant, the enthusiasm of your personnel may soon falter. You may even incur a large expense and be perceived as overacting.

Energy Security Plan

The best possible solution to this dilemma would be to formulate a physical plant security plan as part of your overall energy plan. Then, as with the other parts of the overall plan, you would only implement those stepped-up security measures when international energy trends and other security indications warrant it.

All of this is fairly well-established in the nuclear power industry. The level of security at this type of facility has as its primary intention the protection of the public from a radioactive release caused by an act of sabotage. From a purely energy standpoint, this is a curiously lopsided concentration of security emphasis; nuclear power facilities are a small portion of the entire electric power production industry and an even smaller portion of industry in general. The backbone of the electric power production industry are the multitude of fossil fuel-fired generation plants that are scattered all over the country. These facilities, as well as the other heavy industries that consume fossil fuel, lend themselves as easy targets by those who wish to disrupt fossil-fueled production.

TERRORISM AS A POSSIBILITY OR LIKELIHOOD

Before we go further, it may be wise to consider the likelihood of an attack of sabotage or terrorism on your energy supply lines or your facility as an energy generator itself. You may be tempted to dismiss the possibility out-of-hand that it could happen to your plant; at a nuclear plant, sure, but not my little obscure manufacturing venture somewhere in the "piney woods." But not many years ago, a group of Arkansas-based neo-Nazis were charged with planning to dump 200 lbs of pure cyanide into the municipal water supplies of Chicago, New York, and Washington, DC as part of a campaign to overthrow the U.S. government. No, this was not directed against energy supplies, but they *were* based in the piney woods. Yes, it's true that terrorist attacks against industrial facilities are declining somewhat while those involving personal injury are on the rise. But that does not diminish the impact on your plant if saboteurs were to hit your power supplies. And "declining" does not mean "nonexistent." In fact, guerrillas recently dynamited a new oil pipeline in Columbia's Magdalena Valley just as the Columbian state oil company was testing it. This was merely the latest in a long string of terrorist incidents by the National Liberation Army (ELN). In 1988 alone they carried out no less than 58 bombings of the Cano Limon-Covenas pipeline. This pipeline is jointly owned by Colombian, U.S., and other Western oil companies. The ELN has made it clear that one of its major goals is to drive foreign investment out of Colombia. Thus far, Colombia has lost millions of dollars in lost oil revenues.

It has been observed that industrial facilities are decreasing in popularity as targets for terrorist attacks. This is mainly due to a gradual decreasing of world sensitivity to violence which in turn has caused terrorists to target "softer," defenseless victims in increasingly heinous ways. This tends to raise attention levels considered so critical to the perpetrators of international terrorism. Also, the sabotage of an industrial process, although potentially spectacular and immensely destructive to property, takes a rather significant effort to accomplish due to the sophistication of controls and alarms which need to be bypassed, and does not guarantee the expanse of personal injury that a carload of plastic explosives in an open market square does.

However, there have been some near-misses, some attempts that have not received widespread attention. These industrial terrorist attacks occurred in the U.S. directly against energy supply lines. Some of them are as follows:

1983: Arkansas river crossing, major natural gas transmission pipeline; bomb detonated; Covenant, Sword, and Arm of the Lord group
1986: Major natural gas transmission system valve at key intersection; bomb removed by FBI; responsibility unclaimed
1986: Quachita Parish, Louisiana natural gas pipeline being built; small incomplete holes drilled into new sections of pipe for delayed corrosion leakage; responsibility unclaimed
1989: Louisiana propane pipeline; holes drilled into sections of pipe; responsibility unclaimed
1989: Arcadia, Louisiana natural gas pipeline; bomb detonated; responsibility unclaimed
1989: Arcadia, Louisiana natural gas pipeline; bomb removed; responsibility unclaimed
1989: Sterlington, Louisiana natural gas pipeline; bomb removed; responsibility unclaimed

It is neither correct nor prudent to assume the attitude that any country is immune to the actions of saboteurs or industrial terrorists.

In order to formulate a defense against industrial terrorism, it is important to try to think like a terrorist. Most individuals have no idea how a terrorist thinks, which points once again to the need to involve a security consultant early on. Most terrorist organizations don't have much economic resources. So, they want to cause the most amount of damage with the least amount of cost. And the need to create attention is perceived as critical to the effectiveness of a terrorist's actions. These individuals are relentlessly focused upon the perceived legitimacy of their actions. Terrorists tend to have a military ideology which they use as a creed. They are typically individualistic in their thoughts and behavior patterns.

One of your primary concerns is the sustaining of adequate supplies of finished or raw energy to your facility in order to maintain production. Although the threat to a *physical system* due to terrorism can be dismissed as unlikely, you and your employees are more likely to face this threat. It is difficult, if not impossible, to operate a production facility without a technical staff who enjoy an adequate sense of security. This goes for you and your personal security, too. Additionally, just because an industrial attack by terrorists is unlikely, it nonetheless remains a possibility. A single, disgruntled employee has the access, knowledge, and motive to significantly increase the likelihood of physical system damage.

TERRORISM—WHAT WE KNOW

Some semblance of security already exists at most facilities. This is mainly to protect companies from theft, vandalism, loss of property, and to ensure personnel safety. What usually happens at a plant is a gradual relaxation of security measures to a level which allows the guard to stay in his shack and wave at people entering and leaving the plant. This is unfortunate because nominal gate security does not come close to protecting your facility from the rather destructive phenomenon of industrial terrorism which we see occasionally, particularly in developing nations.

In order to adequately assess the likelihood of a threat against the energy supply systems to your plant, we need to look at what drives terrorism. Let's look at the types of terrorists involved. Let's take a look at the names they've given themselves and where they're headquartered. Remember, this is all part of "assessing the threat," i.e., knowing your enemy.

There are many definitions of terrorism. Basically, it is premeditated, politically motivated violence, perpetrated against noncombatant targets by subnational groups or clandestine state agents, usually to influence an audience. It is intended to cause a great psychological impact upon its target.

Terrorism takes six basic forms: bombings, assassinations, armed assaults, kidnapping, barricade and hostage situations, and hijackings. Bombings are the most common.

Actually, a terrorist can be any individual that is willing to use violence to bring attention to that individual's cause. Former or present employees, psychologically unbalanced individuals, ultra-right or -left groups of local citizens: all are capable of terrorist activity. What marks them as terrorists is that they direct their violence against noncombatants with the goal of terrorizing a wider audience than the immediate victims, thereby attempting to gain the political influence they desire over the larger audience.

Often, terrorist groups are relatively small, tightly knit, politically homogeneous units that are rather incapable of developing popular support for their radical positions and so have turned to violence to gain influence. Another variant of terrorism is conducted within the context of insurgence. In this case, wide-scale revolts against the established government are conducted by paramilitary or guerrilla forces operating within the boundaries of the country under siege. The insurgent group usually has a terrorist component seeking to undermine the government's credibility, legitimacy, and public support.

Probably the variation of terrorist that poses the greatest challenge to industries connected with the United States or other democratic countries is "state-sponsored" terrorism. This is terrorism that enjoys direct sponsorship from sovereign states for their destructive activities. This sponsorship intensifies the degree of horror or destruction that these people wreak and widens their radius of influence considerably.

In order to get an idea of the wide breadth of possibilities, we need to look at *current* known terrorist groups and the regions they operate in.

Middle East

Middle Eastern terrorism is driven by such things as the issue of a Palestinian homeland, Israel's existence and policies, and regional jockeying for power among Arab states. The following organizations are known to exist:

Table 4–1 **Middle Eastern Terrorist Organizations**

Name	*Operational Base(s)*	*Target Nations*
Abu Nidal	Libya, Lebanon	USA, United Kindgom, France, Israel, non-sponsor Arab States
Al-Fatah	Tunisia, Lebanon	Israel, Jordan
Hawari Group	Middle East, Europe	Syria, USA
Armenian Secret Army for the Liberation of Armenia	Lebanon, Europe, USA	Turkey, France, USA, Switzerland
Democratic Front for the Liberation of Palestine	Syria, Lebanon	Israel
Force 17	Lebanon, Europe	Israel
Hizballah (Party of God, Islamic Jihad)	Lebanon, Europe, Africa	USA, Israel

Table 4–1 Middle Eastern Terrorist Organizations (continued)

Name	*Operational Base(s)*	*Target Nations*
Justice Commandos of the Armenian Genocide (Armenian Revolutionary Army)	Europe, USA, Canada, Middle East	Turkey
Lebanese Armed Revolutionary Faction	Lebanon, Europe	USA
Palestine Liberation Front	Iraq	USA, Israel
Palestine Liberation Organization	Tunis, Middle East	Israel
Popular Front for the Liberation of Palestine	Syria, Lebanon, Europe	Israel, moderate Arab states
Popular Front for the Liberation of Palestine—Special Command	Lebanon, Europe	USA, Spain
Popular Struggle Front	Syria, Lebanon	Israel, moderate Arab States, PLO
Sa'iqa	?	?

Europe

Terrorism in **Europe** comes from a multitude of indigenous groups as well as organizations based in the Middle East. We've already addressed ideologies of Middle East groups. The indigenous groups usually espouse a revolutionary philosophy very similar to a 19th century anarchist style. Most are dedicated to overthrowing the existing government but are sketchy about what should take its place. These groups generally select their targets very carefully, opting for those with as much symbolism as they can manage. People and businesses connected to the United States are frequently attacked because the "imperial" symbolism associated with the U.S. is a high-profile matter. U.S. and allied military individuals and facilities are also favorite targets. Some of these groups are "ethnic separatists" and target only specific ethnic groups. These indigenous groups tend to be small in size. The following organizations are known to exist:

Table 4–2 European Terrorist Organizations

Name	*Operational Base(s)*	*Target Nations*
Action Directe	France	France, USA, Israel
Basque Fatherland and Liberty	Spain, France	Spain, France
Communist Combatant Cells	Belgium	NATO allies
Direct Action	?	?
Iraultza	?	?
Irish National Liberation Army	?	?
Kurdish Worker's Party	Turkey	Turkey
October 1st Antifascist Resistance Group	Spain	USA
Popular Forces of 25 April	Portugal	USA
Provisional Irish Republican Army	Great Britain	Great Britain
Red Army Faction	Germany	Germany, USA
Red Army for the Liberation of Catalonia	Spain	USA
Red Brigades	Italy	Italy, USA
17 November	Greece	USA, Greece
Revolutionary People's Struggle	Greece	USA, Greece
Terra Lliure	Spain	Spain

Asia

Terrorism in **Asia** has steadily increased since about 1986. International terrorist incidents increased dramatically from 56 in 1989 to 96 in 1990. This increase is due mostly to activities by Afghan agents in Pakistan and Communist guerrillas in the Philippines. Most of the violence results from ongoing insurgencies, ethnic conflict, and domestic disputes. The greatest threat to U.S. personnel and facilities is in the Philippines, where the New People's Army, the military wing of the Communist Party of the Philippines, is very active. It is not known how their activity level will trend now that the U.S. has closed all of its military installations there. The following organizations are known to exist:

***Table 4–3* Asian Terrorist Organizations**

Name	*Operational Base(s)*	*Target Nations*
Chukaku-ha (Middle-Core Faction)	Japan	Japan, USA
Dal Khalsa	?	?
Dashmesh Regiment	?	?
Japanese Red Army	Lebanon	Japan, USA, Israel, Italy, Spain
Liberation Tigers of Tamil Eelam	?	?
New People's Army	Philippines	Philippines, USA
Sikh Groups	India, Europe, North America	India, Japan

Latin America

Since about 1979, terrorism has been steadily increasing in **Latin America**. In fact, the number of international terrorist incidents there rose to 162 in 1990, higher than any other region of the world. About two-thirds of all anti-U.S. attacks in the world took place in Latin America. The fuel for this apparently comes from Cuba's enthusiasm for promoting revolutionary violence and involvement by Nicaraguan Sandinistas. Latin American terrorist groups tend to be highly nationalistic.

Five of the six incidents in Bolivia were anti-U.S. in nature. One group abducted the Bolivian president of the Coca-Cola Company and murdered him. The country of Chile topped the list of nations worldwide for anti-U.S. attacks in 1990, totalling 62. Most attacks were executed against U.S.-related property, not individuals. Columbia's democratic government is opposed by active leftist guerilla groups, rightwing paramilitary groups, and well-financed narcotics-trafficking organizations.

Probably the most significant energy-related attacks by a terrorist group occurred in El Salvador where, in the latter part of 1990, the Farabundo Marti National Liberation Front conducted more than 100 attacks on the electrical power grid and hydroelectric plants. These attacks caused more than $10 million in damages. In December of that year terrorists disabled 10% of the country's telephone system. Three major Guatemalan guerrilla groups struck many economic and nonmilitary targets such as bridges, powerlines, telephone systems, and government road repair facilities. Also, since it was finished in 1986, a major oil pipeline in Colombia has been the target of numerous acts of terrorist activity, the most recent of which occurred in late 1992. The following organizations are known to exist:

Table 4–4 **Latin American Terrorist Organizations**

Name	*Operational Base(s)*	*Target Nations*
Central American Revolutionary Worker's Party	El Salvador, Costa Rica, Honduras Guatemala, Nicaragua	USA
Clara Elizabeth Ramirez Front	El Salvador	USA
Farabundo Marti National Liberation Front	El Salvador, Honduras	El Salvador, USA
Manuel Rodriguez Patriotic Front	Chile	USA, Chile
Movement of the Revolutionary Left	Chile	Chile
Movement of 19 April (M–19)	Colombia	USA, Chile, others
National Liberation Army	Colombia	USA (esp. oil indus.)
People's Liberation Army	Colombia	USA
Popular Revolutionary Forces-Lorenzo Zelays	Honduras	USA

Table 4–4 Latin American Terrorist Organizations (continued)

Name	*Operational Base(s)*	*Target Nations*
Revolutionary Armed Forces of Colombia	Colombia	Colombia, USA
Sendero Luminoso (Shining Path)	Peru	Peru, USA, China, Russia
Tupac Amaru Revolutionary Movement	Peru	USA

Africa

On an international scale, terrorism in **Africa** generally presents a low-level threat. The greatest threats come from Libya's intelligence organization, the Abu Nidal organization (see Middle East), and domestic terrorist activity by insurgents against their own governments (which can be very significant). The following organizations are known to exist:

Table 4–5 African Terrorist Organizations

Name	*Operational Base(s)*	*Target Nations*
African National Congress	Zambia, Tanzania, Angola	Nonspecific
Mozambican National Resistance	Mozambique, Zimbabwe, Malawi, Zambia	Mozambique

Types of Terrorists

Just by the sheer number of *known* organizations, you certainly could form the opinion that there exists a possibility that a disruption to your plant's energy supply will occur due to an intentional action.

It would be impractical to make an industrial facility an impenetrable fortress. The amount of security you provide your plant will depend on many things. One of the very first considerations you should have is, "Who is my enemy?" You need to analyze the type of individual who might want to breach your perimeter or cut off your energy supplies elsewhere. Some of these people are:

a) a single, disgruntled employee
   - authorized access to plant interior
   - familiarity with facility
b) a single, mentally unstable person or fanatic
   - no authorized access to plant interior
   - questionable familiarity with facility
c) a small group of saboteurs, or industrial terrorists
   - no authorized access to plant interior
   - questionable familiarity with facility
d) an uncontrolled, unorganized mob

Security Engineering

Suffice it to say that the "human threat" is real, all over the world. Even the smallest airports that are served by commercial airlines are required by law to have a certain degree of anti-terror security. Truly effective security for your energy supply lines against the human threat requires a detailed analysis of assets, threats, vulnerabilities, and protective measures. This analysis is called **security engineering**. And it is inextricably linked to the analysis of external energy streams powering your facility. This analysis, described in detail in Chapter 3, involves the technique of **mathematical risk analysis** for ferreting out the choke points and bottlenecks which could spell disaster for you later on. There has been a lot of studying, publishing, stockpiling, and practicing on the part of nations, particular the global superpowers, for the sake of security engineering for a wartime posture. But there is a significant dearth of experience in engineering against the human threat, particularly terrorism.

Choke Points

After you have attempted to identify the types of potential enemies you are facing, it is time to compile a list of your vulnerabilities or energy choke points. These are locations, items of equipment, or facilities that are major energy bottlenecks for your plant. Some examples are:

- Electric substations
- Liquid fuel pumps
- Natural gas compressors

- Electric generators
- Motors, engines
- Overhead power and communication lines
- Railroads
- Conveyor belts
- Barge docks
- Gas, electric, and water meters
- Manholes
- Exposed pipelines and valves
- Wellheads
- Water and waste treatment plants
- Aboveground storage tanks

The types of security measures you should already have in place would include such things as perimeter fencing topped with three-strand barbed wire, area lighting, and a main gate guard shack manned around the clock. Depending upon the likelihood of aggression against your facility, you may want to take extra precautions, some of which are nominal in cost, others of which are not so cheap.

Energy Security Engineering

One of the best courses of action you can take regarding beefing up your physical security is to hire a security consultant. These people can analyze your current security posture relative to your deficiencies. They are abreast of the current technology that can rectify those deficiencies. If you engage a security engineering consultant, make sure that he or she is completely familiar with the U.S. Army Corps of Engineers' manual, "The Security Engineering Manual," written by the Omaha District of the Corps. This is the most comprehensive and focused writing on this subject that you can get. Although a great deal of design data and criteria has been amassed to provide adequate security engineering of facilities in a wartime environment, there is a lack of experience with noncombatant (peacetime) security against a terrorist adversary. The Army manual was assembled to address that lack.

Security engineering is a stepped process that is best performed in the team environment. Ordinarily, is performed in six steps:

1. Identify characteristics of your facility
2. Identify your assets
3. Define the threat
4. Determine your vulnerability
5. Design a protective system
6. Construct the system

This is a *sequential* process; that is, don't skip any steps or rearrange them if you want to optimize your effort.

Step 1: Identify the Characteristics of Your Facility

Some pertinent characteristics that affect your plant's vulnerability would be such things as its proximity to political boundaries and jurisdictions, the type of locale (rural, industrial, commercial, urban, etc.), access routes, natural terrain, and bodies of water.

Step 2: Identify Your Assets

A pivotal piece of input is to assign a value to certain resources, processes, and energy supply streams so that you can get an idea of how much of an investment in these new security measures you want to make. This assessment will usually center on the value of human life, ease of equipment replacement, or maintaining production. Also, remember that assets can be people, material, equipment, information, functions, or operations.

As is usually the case when faced with the need for some sort of prioritization, the **risk analysis process** (Chapter 3) will prove to be invaluable in your assignment of values. These values are identical to the Process Criticality Factors (PCFs) described in Chapter 6.

Step 3: Define the Threat

This step is the *core* of effective security engineering. It is not adequate to merely list the major terrorist groups who could threaten you, much as we did in the beginning of this chapter. The threat must be defined in terms that can be used by a facility design engineer to establish protective measures. One such characteristic needed to begin planning a defense is the attack mode the aggressor may use. There are six common ones: vehicular, exterior, standoff, pedicular (by foot), and chemical/biological.

Another descriptive is the degree of accessibility the aggressor has to the interior of your plant. A single, disgruntled employee poses one of the most difficult threats to overcome from a security standpoint: this individual usually has access to virtually all areas of most ordinary manufacturing facilities. And one of the most effective defenses against this is thorough pre-employment screening, including a psychological analysis, all of which puts you at the risk of being accused of unequal opportunity employment. Another method of defense is to use the observation skills of your supervisory personnel to alert you to potential changes in an employee's attitude.

That way, a lot of grief can be avoided by defusing the problem brewing. Also, it is a good idea to conduct a thorough exit interview of a departing employee to give that person an opportunity to air out any grievances.

The use of intelligence can be a help in defining the threat. In this sense, "intelligence" is defined as the gaining of information via conventional or unconventional means to prevent an act of industrial terrorism. Interviews with the general population, the media, the government, or any public source constitute "conventional" intelligence gathering. "Unconventional" intelligence gathering consists of using such sources as governmental agencies (FBI, CIA, etc.), electronic eavesdropping, interviewing fellow employees, etc. utilizing sources beyond normal legal means. Unconventional intelligence is commonly used by terrorist organizations to gain information about your facility. Bear in mind that corporations in the United States are legally limited to the use of conventional means of gathering information. Multinational corporations employ it abroad where it is not prohibited. Security intelligence should be employed as early as possible during the design stage of a greenfield facility to provide for as defensible an installation as possible.

The military is attuned to the establishment of fortifications in areas that can be easily defended. However, industrial installations are not usually located with defensibility in mind. Rather, accessibility to utilities, fuels, and raw products tend to dictate the siting of manufacturing plants. What this does is make these necessities, the gas lines, the water treatment systems, the electrical transmission lines, etc., key targets for industrial terrorist activities.

Exacerbating this, some terrorist groups, as mentioned before, employ unconventional intelligence gathering to study corporate structures to determine where the greatest damage to the system can be inflicted. Community Right-to-Know laws, originally intended to protect the general public from potential harm resulting from the handling of dangerous substances by an industrial plant, can be used against these facilities by terrorist groups by allowing them basic security information on the handling, storage, and processing of materials inside the fence. You should be aware of this and use your right of confidentiality to protect information about sensitive areas and procedures.

It is disquieting to note how easily some potentially dangerous actions have been performed by nonterrorist individuals such as high school and college students, who have written school papers on weapons construction, computer network "hacking," process disruptions, and who in some cases have carried them out. All of this was made possible using information available at the local library. You should be aware of the possibility of information misuse.

Another realm of unconventional intelligence by industrial terrorists takes place in the local dumpster used by the plant. These folks have been known to dig through garbage to gain intelligence on corporate operations, personnel, and security. Although shredders can significantly hamper this type of espionage, it is not a foolproof destruction method, as was discovered in Vietnam.

Ordinarily, industrial companies resemble the cumbersome bureaucracies of governments, especially when reacting to the attacks of terrorists. Quick and decisive reaction is the exception, not the norm. This works against the plant. Autocratic leadership usually results in rather immediate, well-defined responses to acts of violence directed toward an industrial facility, similar to the response of a totalitarian government. The significance of this type of response is evidenced by comparing the number of terrorist actions taken against the United States vs. against the former Soviet Union.

Step 4: Determine Your Vulnerability

Three factors work simultaneously to map out your plant's vulnerability(ies). They are: 1) the nature of the asset, 2) the goal of the aggressor, and what protective measures are in place *now*. The nature of the asset or energy stream says a lot about the vulnerability of the plant: a computer control system is more sensitive to an electric power interruption, a misplaced crowbar in its side, or the interruption of the cooling system; any single one of these can bring down the power house and thus flatten the facility's production. Conversely, it is not very vulnerable to the effects of blast overpressure or glass shards from a bomb exploding. So, the focus or goal of the threat needs to determine the performance criteria of the protective system you employ. And this is likely as not to allow certain facility failure modes if these are not mission-critical or life-threatening. Some of the most effective protective measures simply deny access of the asset to the threat.

Step 5: Design a Protective System

To do this step thoroughly, it should be done in three phases. In Phase I, protective measures are assigned to each of the assets identified in Step 2 for each of the threats defined in Step 3. Phase II is a system integration of all of the protective measures identified into as common a system as possible for all assets against all threats.

Phase III is a value analysis wherein the Phase II system is examined for cost return. The prioritization of assets performed in Step 2 which yielded Process Criticality Factors lends itself handily to the Phase III value analysis. By definition, the PCFs are assigned based upon how critical to production a process or piece of equipment is. Generally speaking, the higher the PCF, the more deserving of being included in an energy security protective system.

Protective systems either function in a *detection* mode or a *defensive* mode. Detection is rather self-explanatory, and includes closed-circuit television (CCTV), hidden microphones, motion detectors of various types, watch dogs, etc. The defensive mode either shields the asset from the threat, hinders the advance of the threat toward the asset, or aids law enforcement in defending against the threat.

Detection Systems

Multitudes of perimeter intrusion detection devices available, which are constantly being improved and becoming obsolete. The major types are a) microwave, b) electric field fence (EFF), c) video motion detection, d) stress/vibration, e) seismic, f) ferrous metal, g) ultrasonics, and h) infrared. As varied as they are, so are they priced and so are they effective. Actually, any of them can be depended upon to function poorly or quite well, depending upon the application. Some have lower initial costs, but have reduced reliability and may be harder to install properly. Also, some have lower false alarm rates than others.

In most cases, a dual application of detection methods is worth the expense. For instance, strategically positioned CCTV can provide assessment of a threat that set off an intrusion device. A CCTV camera scanning an alarm zone can quickly inform security personnel whether or not the intruder is an aggressive raccoon or a human intruder. Such a camera provided with the options to pan, tilt, and zoom can track an intruder's actions while security forces are being dispatched. But even with all of these options, CCTV is generally a poor stand-alone type of security system; it is best used in conjunction with a more passive intrusion detection device. Also, CCTV requires another simple security measure, adequate lighting.

Lighting at sufficient intensities along the plant perimeter and throughout the site is necessary for minimal security. Nuclear power plants are required by law to have a minimum lighting intensity of 0.2 footcandles (fc) along the plant's entire perimeter, covering no less than 20 ft of cleared ground outside and inside the fence. Some other areas need lighting at other levels. For instance, emergency interior lighting requires 3 fc on the floor. Some exterior areas and lighting levels are as follows:

a. Unobscured roadway: 0.5 fc
b. Obscured roadway: 1.0 fc
c. Catwalks, Substations: 2.0 fc
d. Coal unloading platform: 5.0 fc
e. Main entrance: 10.0 fc
f. Fence, open area: 0.2 fc

Sometimes it is not appropriate to light an area continuously. Besides being costly, it removes the cover of darkness from responding security forces while they are moving in on some intruder. In situations like this, one method would be to have all security-related lighting turned off until a zone detection alarm goes off. An appropriate subsequent action would be for only those lights able to flood the compromised area be called up by the alarm and turned on. When you are planning your lighting scheme, remember to provide for redundancy and standby power as foils to attempted sabotage.

There are many other considerations that are best left to a security consultant. For instance, if you are in an area that is prone to lightning strikes, this would influence your choice. Several electric field fencing (EFF) zones can be wiped out by merely a proximate lighting strike, while a microwave system would only suffer damage to a single zone. However, if you are surrounded by rough terrain, a microwave system becomes ineffective compared to the EFF system, which has no line-of-sight requirements. Also, Where does the need for CCTV come into all of this? What is cost-effective? It depends upon your facility and the value you place on security for your personnel and production.

Defensive Systems

It was stated before that defensive systems either shield the asset from the threat, hinder the advance of the threat toward the asset, or aid law enforcement in defending against the threat. Because there are usually a significant number of assets to be defended against a multitude of threats, sorting out the final choices can be a cumbersome task. Again, this makes a definite case for retaining the services of a security consultant.

And you can't expect to get much from this person if you just box up what you've put together, drop it off at his or her office, and leave a note that you'll be expecting justified recommendations in a couple of weeks. It really goes without saying (this is in case you may forget) that because of the complexity of prioritized assets, the various probabilities of engaging a wide array of threats, the advantages and disadvantages of each type of detection or prevention system, and the various costs of each type of installation, you'll need to make sure that your consultant will end up knowing more about your entire operation than most of your employees, if you don't want to waste your money.

Some defensive measures are simple yet cost-effective. For instance, the costs of protecting a system from an explosion can become excessive due to the massive differentials in air pressure and the shrapnel being hurled at extreme velocities. The explosive loads decrease by the cube root of the distance of the center of the detonation. So, the most cost-effective anti-explosion defensive measure is couched in distance. The more distance you can keep between the asset and the center of the explosion, the less the asset needs to be designed to withstand the effects of a blast.

Of the different types of attacks that occurred against U.S. targets in 1990, 63% were bombings; no other type of incident came close to this. Armed attacks were 13%; arson incidents were 3%; firebombings were 9%; kidnapping was 8%; the remaining 4% were classified as "other." So, protection against bombings has real merit.

Distance being the most effective defensive measure against the most likely type of attack, bombing, it is necessary to provide for this "stand-off" distance, sometimes referred as a **blast zone.** This zone can serve as an **isolation zone** to buffer the asset from other types of attacks than bombings.

This brings us full circle to the **requirements for detection**. Detection is the primary means necessary for establishing the blast zone. An entry-control point (E-CP) into the zone needs to be created as the only way into or out of the zone. This E-CP must be designed such that vehicle searches can be easily carried out. Also, once a vehicle or individual gains entrance through the E-CP, they should not be automatically allowed to leave the plant without undergoing a thorough search. One effective method to prevent unauthorized vehicle egress is to employ a remotely activated hydraulic vehicle barrier. One of these barriers can typically stop a 15,000-lb vehicle travelling at 50 mph.

The perimeter of the blast zone created around the protected asset(s) should be matched to the expected threat. Sometimes an 8-ft chain link fence with three strands of barbed wire isn't nearly enough. You may wish to electrify your fence, install razor tape, or

Figure 4–1 **Examples of perimeter protection**

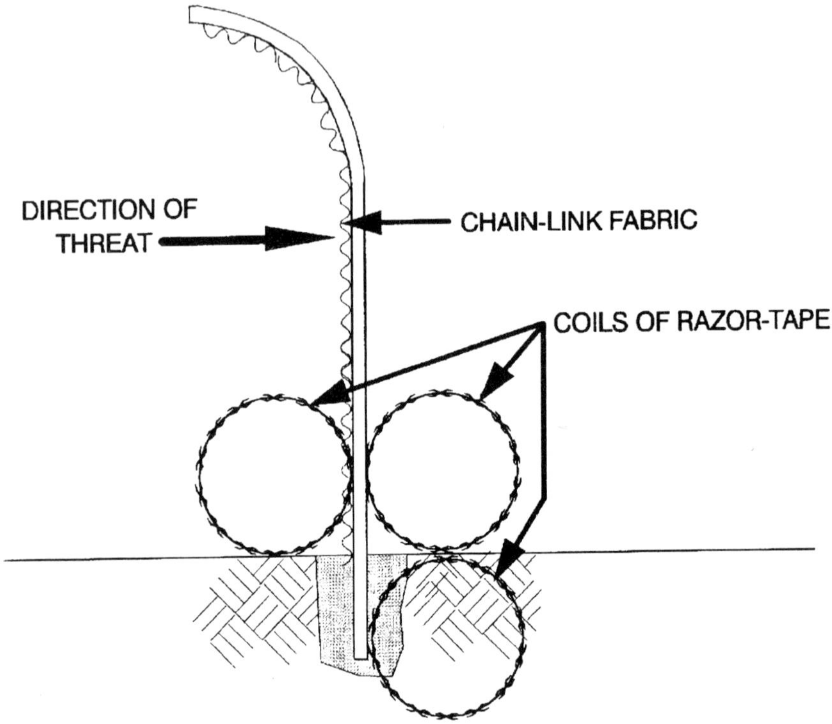

replace your fencing with the outward curved type (Figure 4–1). Also, you might consider the use of patrol dogs along your perimeter as an economical deterrent to entry by unauthorized personnel.

The next most common type of aggressive perpetration is the armed attack. This can be done from a distance with high-powered rifles, mortars, or bazookas, or it can be done from a closer distance to allow the hurling of rocks, explosives, and incendiary devices. The strategy against these types of attacks is to shield the asset using blast hardening or obscuration to foil attempts at visual, acoustic, and electronic eavesdropping. Although effective defensive postures can be built to resist armed attacks, there are no adequate detection methods to warn security personnel, since this type is a "quick-strike" type of aggression.

The other types of aggression popular among industrial terrorists are kidnapping and arson. These require the individual to gain entry into the isolation zone to damage the asset. The perpetrator gains access by 1) forced entry, 2) covert entry or surreptitious entry, and 3) authorization. In any of these three cases, the **strategy of defense** is to detect the aggression and delay its progress toward the asset until such time as local law enforcement can successfully engage and neutralize the threat.

To defend against forced entry, typical safeguards are hardened construction and intrusion detection, which were described above for perimeter security.

Covert entry can be hampered by using such mundane things as limited-edition keys and standardizing issuing procedures for them. All access doors should have an alarm sensing device to alert security personnel when the door is being opened. Other more sophisticated controls can verify a person's identity using an eye retina scanner, a voice comparator, or a fingerprint analyzer and subsequently grant an authorized person access.

The options to guard against industrial terrorism by the authorized individual are limited in number and by laws protecting the privacy of the individual. Weapons and explosives can be detected by guard dogs, or by guards using metal detectors, fluoroscopes, and vapor trace analyzers.

Trained guards, either armed or unarmed, function as an integral portion of your overall strategy of access denial to the industrial terrorist. Humans make better decisions than sensors and cameras, because visual acknowledgment of a situation is critical to countering terrorism. No matter how sophisticated a detection system you installed, the "rubber hits the road" by sending someone to physically check it out. However, to frame the proper response to a security breach, some sort of procedure needs to be in place to assure that the degree of response is adequate without being overreactive. These procedures need to be agreed upon and formalized by all parties involved, and personnel need to be trained on them.

Physical security methods are available in a wide array. Which ones you actually adopt will depend upon which energy supply lines are most vulnerable to a human-caused event, how critical to your continued production the energy stream is, how likely a target for sabotage or industrial terrorism you or your energy supplies are, the degree of simultaneous protection you can give your people and your operations using the same security systems, and your budget.

If you are instituting a completely new or an upgrade to an existing security system, you should take a phased-in approach to getting it placed in service. This will fend off tendencies for energy security to become lax because you'd be introducing something new at different points over a period of time. Your employees' attitude toward energy security and physical security in general would remain crisp, which is essential to responding to a security situation adequately; the best time for an industrial terrorist to strike is when attitudes toward energy security are lax and guards are down.

There is not much you can do to contibute to the physical security of the energy infrastructures that keep your plant running. You can, however, insist that you're made aware of the efforts your energy suppliers have taken against the potential of sabotage or industrial terrorism. Most electric power companies and gas transmission companies are reluctant to share this information except to assure you that they are at least as interested in precluding damage to their systems as you are. If they can't sell their energy, they suffer just like their customers who can't sell because they can't produce or operate. If you are not satisfied with this reasoning, it may take insistence from your upper management to their upper management to retrieve this information.

The resolve and technical competence of saboteurs and industrial terrorists should never be underestimated, nor should the likelihood of your facility being targeted for attack. When you couple that with how intimately your ability to maintain operations is tied to the well-being of some rather vulnerable energy supply infrastructures, your need to be concerned about the physical security of your plant and its energy supplies becomes much clearer.

5

The Flexible Power Option

PRELUDE

Since it began operations, a medium-sized southeastern paper mill enjoyed an enviable position relative to its energy supplies. Its primary source of electricity was an intertie with the local power company through a 138kv substation. It also had a set of boilers and two steam turbines. But it depended upon the power company for most of its electrical needs. The utility was enjoying the benefits of cheap, long-term coal contracts.

A recently hired engineer worked her way up the ranks of the mill management until she was in a position that allowed her influence regarding powering the entire facility. For the five years she'd been at the mill, she had preached caution toward what she viewed as a significant dependence upon the transmission intertie with the power company for electric power. This power afforded the company a healthy product cost advantage due to the cheap energy costs, so her message was not embraced with ease or enthusiasm. But the point she hammered home was their vulnerability to a power supply interruption.

Ever so slowly, the persistent engineer was able to move the behemoth of a company to invest in energy "insurance." She began by surrounding herself with an advisory group of people who represented a large cross section of experience, skill, and expertise. They gradually worked together and gelled into a team and began putting together a string of "victories," successful projects which added to their momentum. They began to make things happen to overcome what they perceived as the mill's overdependence on cheap purchased power.

One of the "victories" they claimed was modifying the hog fuel boiler to burn No. 2 and No. 6 oil as well as natural gas. They installed a "hog" to chip up unconventional wood wastes such as pallets and branches, etc. They began to purchase and burn the woody trash that is usually left on the forest floor after timber cutting. They installed piping that would allow them to mix fuel oil with "black liquor" for burning in their recovery boiler (specific to the Kraft pulping process for paper). A by-product of pulping softwood by means of the Kraft process is turpentine, which is usually accumulated and sold. The team had piping installed which would allow this fuel to blend with the No. 6 oil for viscosity control and improved lighting characteristics.

They went even farther than that. And this really paid off as you'll see later. They influenced their office of mineral holdings to pursue negotiations with an oil company to develop their modest gas reserves on their timber property.

It took many months for the major redundancies to be operational, but by late summer 1983, the mill was ready to face almost any energy supply disruption.

That was fortunate, for by mid-September, a litany of potential problems were lined up against this paper mill in eastern Mississippi. As you can imagine, some contingencies occur quickly, others more slowly. Some contingencies are brought on man and his civilization, others escape the hand of man entirely.

What this mill faced had grown slowly but steadily. Over the course of several months, the entire region surrounding the mill had stood baking under a relentless summer sun, driving daily temperatures well into the 100s and leaving pitiful little chance for moisture to condense into rain clouds. By the end of September, the drought was solidly entrenched.

Under normal conditions, the paper mill would use treated river water for once-through cooling of its many internal processes. As the drought dug in, however, the river level plummeted and the paltry remaining flow of water got very warm in the summer sun. The mill had to construct emergency cascade cooling towers and repipe water so that it would recirculate. Ordinarily, the powerhouse at the facility generated around 25% of the electrical power of the mill, although turbogenerator capacity was about 55%. Additional power was required due to recirculating cooling water, increased refrigeration demand, and fans on the cooling towers. Obviously, they needed to pick up their self-generated fraction.

There is probably no other industrial process that lends itself as readily to multifueling as the Kraft wood pulping process. This mill was typical. Specialty boilers were used to recover the fuel value in the lignin and polysaccharides resulting from the "cook" of the wood chips. Natural gas was also piped to the chemical recovery boilers, mainly for startup but also for additional firing capacity. As mentioned before, the foresighted repowering team had arranged for fuel oil lines to be run to these units, too.

The other boilers were configured to burn bark, waste wood chips, coal, No. 6 oil, natural gas, and fiber sludge from the wastewater treatment process. They had even successfully burned petroleum coke and sugarcane bagasse (postextraction residue) in these combination boilers. As it became economical, they stockpiled coke for this purpose; they could get bagasse fairly easily.

Besides the bark, they had an arrangement with the pulpwood harvesters whereby they would even purchase truckloads of logging remnants, the limbs and leaves that would ordinarily be left behind on the forest floor. They chipped this stuff up and burned it with the bark. The variety of fuel they were using by the time they completed the multifueling program was absolutely amazing.

As the drought had intensified, so had the efforts to keep the mill in production. Additional hog fuel was stockpiled. Efforts were stepped up to locate additional cheap coke from Louisiana and Texas, which they stockpiled. They also arranged to receive No. 6 fuel oil as a royalty "in kind" for the natural gas being produced on portions of the company's land. They kept the storage tanks full. This arrangement allowed them to install an additional tank that doubled their original storage with eight days of capacity. Well, the repowering team was pleased that they seemed to be holding their own....

They did not foresee that they'd be the target for an attack.

The enemy they would face was totally devoid of compassion. But he was not nameless, nor was he silent. His stalking was relentless, and furious. He had many names, most of them unkind. *"El Diablo esta despierto"* (the Devil is awake), cursed one Mexican-American truck driver who regularly hauled Texas coke into the plant.

And the enemy came; he crept doggedly toward the paper mill that was hanging desperately on to its production in the midst of the worst drought in decades. His advance could be seen for almost a hundred miles. He sent gray and white waves of smoke plumes like the awesome standards of approaching battle.

As the huge forest fire cut its path toward the mill, a decision to continue production was made on a daily basis. The news media held the fire as the top story for about a week, announcing successes and setbacks of the U.S. Forest Service firefighting effort. Meanwhile, supplies of raw material and hog fuel began to show signs of strain.

Plant personnel set about making preparations for the physical onslaught of the fire, which had the potential for virtually surrounding the whole facility for agonizing hours. The sobering reality of the situation would be brought home to the plant much sooner than anyone had imagined. But at least the possibility had been anticipated.

The wind changed course and within a matter of about two hours the front of the blaze had eaten its way to the right-of-way through which the main electrical power line to the mill ran. The intense heat caused an exaggerated sag in the large cables. But it was the soot, smoke, and flying debris encrusting the ceramic insulators that allowed the high voltage to arc to the ground. The relays kicked out the circuit breakers at the mill and killed all incoming electric power. But the mill had been working to decrease its dependence upon this purchased power.

The prior planning of the in-house utilities group also included putting together a standard operating procedure for the orderly, sequential dropping of prioritized electrical loads. When the interruption occurred, execution of the procedure was initiated. Critical production and environmental protection loads were able to remain on line, while less critical loads were scaled back or cut off.

It would be unreasonable to pretend that full production went uninterrupted. They had to scale back some production. *But they did not have to shut down completely.* They stayed in environmental compliance (which saved them a mountain of headache-inducing paperwork and fines). They kept a great deal of product coming out the front gate. They kept people gainfully employed without interruption, including vendors and suppliers.

Their ability to use a wide variety of fuels in various mixes for boiler fuels was the greatest success factor in their favor. Had the mill demonstrated any vulnerability in this regard, it would have to shut down entirely and indefinitely to wait out this natural disaster. Multifueling turned out to be their "ace in the hole" and they played it well.

El Diablo que esta despierto was finally put to sleep by an intense weather system which brought a salvational change in wind direction and three inches of steady rain. Within two weeks after emergency procedures were begun, the mill was operating at its normal fuel mix, and at normal production levels. This mill was unique. It was a survivor due to level-headed planning using the options at hand.

Fueling Options

This illustrates one of the most capital-intensive, yet very effective methods of securing energy for your plant: Give it as many options on fuel or other energy sources as you can afford. Although it is physically possible to enjoy a wide variety of these options, capital considerations usually limit exploitation to duplication of fueling. More often than not, this appears to be adequate and is certainly better than having only a single source of primary process energy. You have the laws of chance in your favor; it is highly unlikely that more than one major energy source will be disrupted or become overpriced at the same time.

The concept of energy security can go beyond the physical assurance of continued operations despite supply disruptions. It also includes the indirect economic considerations which underlie everything you do at your plant irrespective of your physical fuel needs. In other words, multifueling uniquely impacts both the physical and economical ability of your plant to run.

Fuel Shopping

The physical is obvious: since the supply of primary fuel is subject to disruption, and production with it, you install provisions to switch to a secondary fuel. Thus, you've enhanced your ability to maintain production. The economic considerations are a bit more subtle. Multifueling has the additional benefit of allowing you to take advantage of fluctuating fuel prices among the different raw energy streams into your plant. So, you don't necessarily need a supply interruption to justify it. Similarly, it also allows you the luxury of being able to play one fuel supplier against the other with real credibility! Now, that's worth some *satisfaction* besides cold cash savings. Again, you're assured cost effective productivity irrespective of curtailment insurance. Having multiple energy options is a very effective energy security measure, but because it is very capital-intensive, *you will need to actually employ these options for the sake of cost-effectiveness.* Otherwise, you will lose in a couple of ways. First, you'll have capital tied up in idle facilities. Secondly, in the absence of switching, the facilities and equipment which allow you this option are prone to neglect and need to be maintained and periodically checked for proper operation.

And there's no law against smart planning, either, when it comes to fuel choices. Just because garbage is cheap and available doesn't mean that it's the fuel for your future. Of course, it could be, *if* you're building a steam/electric power plant that is near a major metropolitan area.

Energy Cost VS. Shutdown Costs

Redundancy, duplication, and diversification in energy supply come with additional costs which you must weigh against the costs of shutting production down. Most of the time, production shutdown is so costly that measures to stay on line can be easily justified economically. So, how probable is a loss of a particular energy source, either because it is physically unavailable or is priced totally out of reason?

Application of Risk Analysis

One way you can gain some insight into this probability is to perform **risk analysis** (see Chapter 3) upon each of the external energy inputs into your plant. Certainly, your access to necessary information on utilities outside your plant will be limited, so your analysis of risk will not be as rigorous as that you will be able to perform on your internal components of demand. But remember that you are just looking for some sort of *relative* energy supply vulnerability. Risk analysis should yield a priority list of critical energy supply components which can give you direction as to where best to focus your attention (and capital) for multifueling.

Another factor you have to take into consideration is the expense you face in adding the ability to use a secondary fuel. Some things are possible for minor investments in capital such as burner-tip retrofits to fire natural gas in an oil-fired boiler; or the installation of No. 6-oil flexible piping and nozzles to stick into the black liquor gun ports in a Kraft mill recovery boiler; or changing your duct burners in the exhaust of a gas turbine to burn No. 6 oil besides gas. Initially, look at what is possible short-term. Usually, that turns out to be a low-cost option, too.

Long–Term Planning

However, always keep your sights on the horizon and gradually begin moving your facility toward other, more long-term, solutions. These would be "big ticket" items like a major electrical-generating unit, an additional steam-generating unit, a cooling tower, a feedwater heater, or your own tie to another electric utility or gas main line. These are true strategic moves that require advanced budgeting after much study. But just because possible benefits will not accrue until sometime in the distant future is no reason to postpone the strategic planning process. The study of the optimum process configuration you are entertaining will take months, months you'll later wish you'd have finished with already. These strategic plans require significant capital to put into place. But that could be offset by the benefits of enabling your plant to use a cheaper energy source.

Using a cheaper energy source could be further optimized by using the multifuel capacity of your installation to play the suppliers of each fuel against each other to work your fuel prices down.

Backup Power

Another aspect of multipowering is the installation of backup power systems for electrical and mechanical drives. This is most commonly implemented on a limited scale for the energizing of emergency circuits and machines and is rarely sized large enough to power the entire load of a facility. What happens in the usual case is that a limited potential is envisioned for the backup system. One reason for this is that the backup power system is assumed to be idle most of the time, during which it essentially locks up capital. It just makes sense to lock up as little capital as possible.

When backup power is envisioned in this light, the equipment is sized in accordance with the minimum demand necessary for essential processes, especially those which are essential to assure safety for personnel. And in most cases, this is sound planning. Most industries would have a hard time generating on-site electric power that would be cheaper than the power company serving them. However, a compromise may be possible with the purchase of less expensive "off-peak" power. Then, armed with this less expensive power and an on-site backup power source, you may be able to configure your operation to use only "off-peak" power for your main manufacturing needs and severely curtailing manufacturing, and hence, constraining your load during "on-peak" times. An on-site backup power source could be used during peak periods to maintain critical control and instrumentation systems while you curtail the main portion of your operations. This would be a good time to schedule maintenance on the units that are shut down. It would also allow you to operate the backup system regularly and minimize maintenance due to neglect.

Strategic Plan

All of this, as complicated as it may seem, needs to be set forth in some written form so that it becomes a part of your entire energy strategy. The proper place to put all of this in writing is in your Strategic Energy Plan so that your entire facility will be operated in a more singular direction.

What Winston Churchill said about oil almost 80 years ago on the eve of World War I still speaks wisdom today: "On no one quality, on no one process, on no one country, on no one route and on no one field must we be dependent.... Safety and certainty in oil lie in variety and variety alone."

6

The Energy Contingency Plan

PRELUDE

It was early October. Fall was rustling the leaves with the wind. Usually a pleasant time for kids, young moms and new dads. This autumn proved not so pleasant for one such new dad who had just been promoted to a supervisory position in a polymer manufacturing plant. His facility enjoyed prominence as the major employer and model corporate citizen to this medium-sized town in western Indiana.

The town got its water from a single deep well that was powered by a 150-hp semi-enclosed, fan-cooled motor. The location of the well, which was outside the town limits, put it on an electrical distribution feeder that also fed the polymer manufacturing facility. Early voltage problems resulting from starting the well motor and the motors in the polymer plant caused the rural electric co-op to install a bank of voltage regulators between the well and the plant and a bank of capacitors at the well transformers.

This seemed to suffice for a little less than two years. And it would have continued to work had it not been for an incident that rocked the usually harmonious relationship between the town and the plant and nearly brought the facility to its knees.

The new dad-turned-foreman was supervising the graveyard shift at the polymer plant on this late October night when one of his operators informed him that incoming voltage to the plant seemed to be dropping steadily. Just as he got up to take a look, his phone rang and the lights dimmed. On the phone was the power company dispatcher, who was calling to request that the plant drop some of its load; it seemed that one of the voltage regulators between the polymer plant and the water well had overheated and failed.

This foreman knew better than to second guess anything about his new duties, so he immediately deferred to the power supply operator as to what should be shut down first. The reply he received was one he'll hardly forget; it was the longest, blankest stare he thought was humanly possible. Probably the only thing that snapped the operator out of it was the foreman's own look of horror when he realized that things were not going to get better quickly.

The two scampered about the dimly lit control room opening dusty three-ring binders that had not been opened since they were put on the shelves decades ago. To the men's dismay, the only usefulness that these ancient documents served was as a haven for multitudinous silverfish who went scurrying everywhere from each book they dared open. The most useful document they found was a very old shutdown schedule that had been used during the last turnaround years ago. Its faded yellow-brown color made it immensely hard to read in the ever-dimming light.

Lacking anything else, the men decided to follow the same schedule. Unfortunately, this shutdown schedule was not designed to drop large electrical loads from the plant's internal distribution system. It was designed to minimize production downtime and, as such, kept some of the major production equipment on line until very late in the shutdown scheme. These pieces of equipment, unfortunately, were very high users of electric power.

The end result of the men's efforts was a matter of too little, too late. By the time they got around to the bigger electrical machines in the plant, the low voltage on the co-op's feeder tie to the water well had shut down the pump motor.

This would not have been a major problem ... ordinarily. What was extraordinary was the low pH of the water supply that the pump provided to the town. This acidic condition had over the years eaten away at the seat of the check valve at the wellhead causing a siphoning effect to occur each time the pump shut off. The water column in the well was free to flow back into the underground sand formation from where it originated, rotating the pump backwards. This had become, over the years, a recurrence after each motor shutdown. It had gone on unnoticed.

This time, everyone noticed.

About two-thirds of the water column had backflowed through the pump and the motor had developed a respectable reverse rate of rotation, when the polymer plant dropped a significant electric load. The sudden resurgence of near-normal voltage caused the water well motor controller to restart the motor. This was an old system, installed many years ago by a small rural community with a very limited budget. Reverse rotation protection was not installed. The resulting current flow, significantly higher than locked-rotor rating, literally melted the motor windings.

The town was on water rations, courtesy of the Indiana National Guard, for two weeks. The slow response to the town's request for a substitute motor from the polymer plant's process stores didn't help matters, either.

ECP—Energy Contingency Plan

What could have been done here? What integral part of energy security planning was missing at this plant? A written, rehearsed, up-to-date energy contingency plan (ECP). You'll hear this plan stressed over and over again throughout this book because of its importance.

Industrial Energy Contingency

Before we go planning for an energy contingency, it would be good to define what it is. Generally speaking, an **industrial energy contingency** (IEC) *is an unexpected inhibition to the ability of an industrial facility to receive sufficient power to sustain normal production capacity.* It interrupts normal operations and demands an immediate, coordinated response by management.

You have either of two ways to handle a contingency: 1) wait until it happens before you try to put together a methodology to cope, and then implement that methodology in the midst of chaos, or 2) take the initiative prior to a contingency and develop an energy contingency plan. Although the latter choice is the better, it is seldom carried out because managers seldom are held accountable for preparing for occurrences that may or may not happen.

Communication Network

The cornerstone to an excellent energy contingency plan is an effective communication network. Truly effective communications enhance contingency response in three succinct stages: 1) planning and training prior to an actual contingency, 2) facilitating solution-building communications during the crisis, and 3) evaluation of damages and preparation effectiveness after the contingency has passed.

Employees

Of all the groups of people *crucial* to the successful weathering of an energy contingency, the *facility employees* stand in a unique position to help expedite the resolution of the crisis or to exacerbate it. These people, especially the ones closest to the energy-consuming components of your operations, hold the key to energy demand and, hence, the longevity of the contingency. Just as nature abhors a vacuum, employees resist information vacuums with rumors. This barrage of misinformation, besides affecting morale and commensurately, response effectiveness, can actually hinder normal communications flow and grind your response to a slow crawl.

Overall Strategy

Although the subject of this chapter is the formulation of an energy contingency plan, this is *only one step* in securing an adequate energy supply for your plant. True energy security is attainable only in degrees, depending upon how far along your plant is along some preplanned long-range energy strategy. Of all the things that come under the umbrella of sound energy management, compiling an **overall energy strategy** and putting it in writing is the most useful thing you can undertake. The most opportune time to assemble this "facility roadmap" is soon after the results of an energy audit (see Chapter 2) are available. The completion of the long-range energy projects identified during the audit forms the focus of your **strategic action plan**; the completion of the short-range energy projects identified during the audit forms the focus of your **tactical action plan**. The short-range energy projects are used to set your facility up to realize the maximum benefit from the long-range energy projects. And, ultimately, the long-range energy projects are intended to secure for your plant an energy posture that is as isolated from power disruptions as you can economically make it.

The final arrangement of this overall strategy is up to you. A suggested outline to start is as follows:

Table 6–1 **Energy Contingency Plan Outline**

Statement of Purpose
Analysis of Situation
Key Planning Factors
Available Assets
(these could go in this chapter)
Facility Energy Management Structure
Strategic Action Plan (Long-Range Projects)
Tactical Action Plan (Short-Range Projects)
Energy Information Plan
Energy Contingency Plan
(As described in Chapter 6)
Expected Results of Planning
Supporting Documentation

You can add to this or rearrange it somewhat. But when you look at what it is you are trying to accomplish with the overall strategy, you'll probably agree that the items above are the minimum that need to be addressed.

Two major targets of focus, long-range energy projects and short-range energy projects, make up the bulk of your overall facility energy strategy. But equally important are the two other types of planning that are part of the overall strategy, the **Energy Contingency Plan (ECP)** and, imbedded within it, the **Energy Information Plan (EIP)**. The compilation of these plans is the subject of *this* chapter.

When planning a defense using limited resources, you must be very careful to maximize the utilization of these resources. This contingency plan is no different. You should focus more of your resources, be they personnel, project capital, maintenance, etc., upon those remedies and safeguards that will be useful against the power interruptions that are *more likely to occur* than others.

Role of Risk Analysis

By far, the best method to use for examining the likelihood of the loss of an energy source is one which incorporates the concept of **risk analysis** into your operations. As is described and demonstrated in Chapter 3, risk analysis is a mathematical method of analyzing the risk associated with a system or interrelated group of components. This method combines the mathematics of statistical probability and the mathematics of sets to sort out the relationships among associated components arranged in a system.

Although risk analysis can be applied to any system of components whose availabilities can be quantified and whose interrelationships are known, it is especially useful to you as a plant manager in two ways. First, you can use it to examine the risk associated with your dependence upon a particular *external* source of power or raw fuel. Second, you can use it to analyze the availability of energy-generating processes *internal* to the plant. In Chapter 3 the use of risk analysis was demonstrated in the comparison of *external* fuel supply networks. The process is similar for *internal* energy-critical equipment.

Since production curtailment can result from either internally caused or externally caused power supply disruptions, you should analyze the risk associated with both.

A primary consideration when calculating the risk analysis posed by dependence on external sources of energy is how your plant fits into the contingency planning of your major energy suppliers (fuel transmission pipelines, electric utilities, railroads, barge companies, etc.). Your facility is on a priority listing with each input stream and is targeted with shutoff at some point in a crisis.

You need to try to find out what that point is for each supply line. Don't be put off by your supplier; your plant is on a curtailment listing, perhaps not by name but at least generically, by the facilities used to get the energy to you. Press them to reveal what condition(s) would force this curtailment. Your planning will be more meaningful and useful than if you were forced to guess when they'll cut your plant off.

This is part of assessing the threat, and is one of the initial stages of assembling your energy contingency plan. Where you are positioned in your supplier's plans can be as important as 1) your regional posture with respect to other energy sources and 2) your proximity to high-threat areas for terrorist activities, sabotage, or vandalism.

The Plan

At this point, we are ready to assemble the actual energy contingency plan (ECP). The ECP should include a general description of the changes of operations necessary to isolate your plant from a major supply interruption. This portion should briefly state the major capital improvements that you expressed in the Strategic Project section of your overall energy plan, including the need for fuel redundancies.

You will need to compile a listing of all the major energy streams into your plant in accordance with the following:

- Type: e.g., natural gas, electricity, coal, steam, etc.
- Source: e.g., pipeline, three-phase transmission line, substation; include pertinent data such as construction material, capacity, normal operating pressure, voltages, ownership, location with respect to ground level.
- Personal Contact: full name of contact, normal and emergency phone numbers, and address.
- Point of Receipt/Discharge: include maps, metering, and highlighted locations of isolation and cutoff valves; include also one-line diagrams and plan view drawings.

- Flowrates: daily consumption rates (or discharge rates for effluent discharges). Include graphs and tables of averages and fluctuations.
- On-Site Production: total daily capacity of in-house production of this type of energy, stated in total quantities and as a percentage of daily energy usage at the plant. Include btu value of liquid, gaseous, and solid fuels.
- On-Site Storage: total capacity of in-house storage of this type of energy, stated in total quantities and as a percentage of daily energy usage at the plant. Include maps of storage locations and highlighted locations of shutoff and isolation valves along with latest revision date of the maps.
- Backup/Standby Sources: fixed and mobile generating plants; include total rated capacity, fuel tank capacity, type of fuel, daily consumption, time between fill-ups under normal loading, general maintenance requirements checklist
- Distribution Network: plan view maps and one-line diagrams of commodity distribution network, both pipelines and electrical conductors. Include all valves and switches highlighted on the drawings as well as latest revision date.

In addition, you'll need to compile a listing of the major energy-consuming pieces of equipment at your facility such as the following:

- Identity of Equipment: name and benefit to the facility process.
- Shutdown Weight: as determined by Process Criticality and Utility Demand Factors.
- Energy Consumption: type of fuel, voltage of electricity, rates of consumption, percentage of facility demand.
- Location: map showing location and single-line drawings of where equipment is in relation to other loads. Include metering, isolation valves.

Energy Posture Levels

The major contribution and the core of the energy contingency plan is a checklist of actions to perform at different levels of an impending or present energy supply disruption.

This action schedule is laid out according to the energy situation at the plant. An excellent method to organize the energy situa-

Figure 6–1 **Operating Posture Levels**

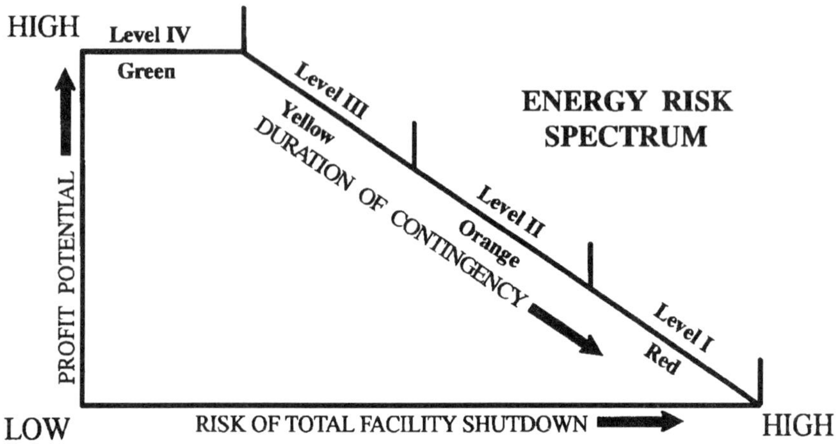

tion at a facility is to use the concept of an Operating Posture Level (OPL) based upon how close the plant is to curtailing production due to an interruption of its energy supply (Fig. 6–1). The following is typical:

Level IV (OPL Code Green): Normal Operations

- No production curtailments exist due to lack of energy supply—*and*
- Fuel inventory levels are normal—*and*
- Capacity margins of energy-critical equipment* are normal

Level III (OPL Code Yellow): Production Curtailment Advisory

- No production curtailments exist due to lack of energy supply—*and*
- Fuel inventory levels fall below one-week supply—*or*
- Capacity margins of energy-critical equipment fall to less than 10%

Level II (OPL Code Orange): Production Erosion Imminent

- Partial production curtailments *mandatory* due to lack of energy supply—*or*
- Fuel inventory levels fall below two-days' supply—*or*
- Capacity margins of energy-critical equipment fall to 0%

Level I (OPL Code Red): Production Impeded

- Production shutdown(s) *exist(s)* due to lack of energy supply—*or*
- Fuel inventory levels are nonexistent—*or*
- Capacity margins of energy-critical equipment fall below 0%**

**Boilers, turbogenerators, pumps, and compressors critical to the internal production of power.

**Critical equipment is operated beyond normal limits (redline) into life-shortening envelope.

For each level the OPL identifies:

1. who performs ...
2. what actions ...
3. with what equipment and materials ...
4. in what sequence.

Another look at the crumb rubber plant from Chapter 3 for an example would be helpful here.

The energy manager had performed a risk analysis on all the incoming energy sources to the plant and had reached the conclusion that there was a higher mathematical probability that the electric power line feeding their substation would fail before the natural gas line would. The risk analysis took into consideration such things as the heavy dependence upon drought-plagued hydroelectric generation, the remote location of the substation, the age of the wooden corner structure that was part of the incoming line and located in a windy slough frequented by sandstorms which were slowly eating away at the poles, the past history of interruptions, etc.

They then used the risk analysis and the relative ranking of failure probabilities to piece together a contingency plan. Of course, this type of plan is very site-specific and theirs was no exception. How will yours look? Let's see....

Energy Demand Shedding

Once the risk analysis of your energy supply system is complete, the next important aspect of your contingency response plan is some sort of internal energy demand-shedding priority listing. Essentially, what you really want to know is in what order the discrete energy-demanding segments of your operation can be shut down during an energy supply curtailment or disruption with a minimum impact upon production and safety. On the surface, this may seem easy to list at first. But when you factor in such relationships among these discrete segments of energy demand as process efficiencies, equipment ages, current market demands, etc., the prioritization becomes clouded at best. It would be helpful to have some sort of objective method of arriving at an emergency shutdown priority that would stand up to logical scrutiny.

Devising a truly objective procedure for prioritization of critical facility components is not as simple, as you can imagine. Why should priorities be assigned objectively? Simply because the relative importance of various sectors of an industrial facility will differ depending upon whom you talk to. Yet each sector will depend on the same energy source for power. So some uniform, objective, and defensible method of ranking units of a process needs to exist. This method should be applied by some sort of authority at the facility, whether it be a composite committee or an individual. Whoever it is should have some autonomy and, of course, the authority to assign these shutdown priorities.

Targeting Power Inputs

Initially, you must decide what you are protecting your plant from. Is it a curtailment of natural gas? An electrical blackout? The failure of your major electric generating unit? A boiler failure? The failure of a major air compressor? The loss of a raw water pump? Although all of these components are essential to the operation of your facility, the likelihood of failure is different for each of them.

To sort out which component offers you the most exposure to a production curtailment, it is necessary to perform a risk analysis upon the systems through which your production receives power. What you end up with is a pretty good picture of which incoming fuel route or power source is putting your operation at the most risk.

When an interruption of one of these fuels or utilities threatens, then major pieces of equipment should be shut down according to a preplanned prioritization based upon two things: how critical a major piece of equipment is to the safe operation of your plant, and how much of that threatened fuel or utility the piece of equipment demands.

Targeting Power Demands

One of the first things you will need to do to start your prioritization process is to list your major in-house utility-consuming operations. Itemize according to your ability to isolate each subprocess and how logically it supports the overall safety, quality, and productivity of your facility. Safety is a rather obvious process characterization. From a quality and productivity standpoint, focus upon the primary mission of your facility, which probably is the production of some sort of physical product(s). Examine how directly each individual energy-consuming operation is linked to your *primary production*. This is an essential first step in selecting those ultimate "undesirable events" for which you will undertake thorough *risk analysis*.

Process Criticality Factor

The rigors of thorough risk analysis will reveal the criticality of your various system components. This is then used to assign a **Process Criticality Factor (PCF)** to each particular operation (component) being prioritized. This gives you some idea of how cost-effective or otherwise desirable it would be to isolate this operational segment from your energy supply. What you are eventually striving for is a roster of these operational segments for your decision making. It may be helpful to group them based upon discrete segments of production, such as a boiler in a bank of boilers or an assembly line for a particular product in a plant assembling other types of products.

Utility Demand Factor

The next aspect you need to consider about an operation is related to utility demand, the rate at which it consumes each of the input utility streams being considered. Considering the electrical requirements of your facility, the term for this rate of consumption is power. Take a look at the sheer magnitude of the power an operation demands, the ease at which the operation can accept a substitute fuel, and the time it would take to repower it with another energy source or type. These considerations are then used to assign a **Utility Demand Factor (UDF)** to the operation. It's entirely conceivable, in fact, it is likely, that a single subprocess or piece of equipment will have a different UDF for each utility stream being considered.

Matrix Prioritization

Both factors are useful when comparing operations against each other for figuring out which ones are candidates for shutdown sooner than others during a contingency involving the particular utility being planned for, be it electricity, a particular fuel, compressed air, or raw water. A simple nomograph, shown in Figure 6–2, shows the

Figure 6–2 **Relationship Among Process Shutdown Factor, Process Criticality Factor, and Utility Demand Factor**

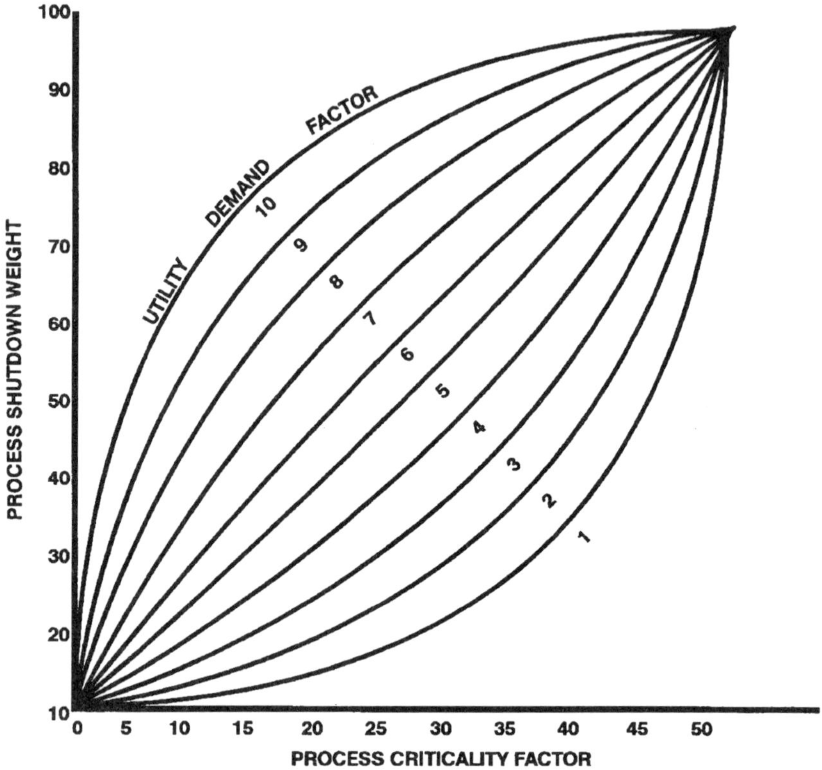

relationship between these two factors and the shutdown priority associated with the subprocess.

With the nomograph, you can observe how different parameters relate to each other when you start prioritizing subprocesses. Using this method of prioritization will add a sense of objectivity to your task, which is essential to cutting through the operational "politics" usually found in an industrial facility. That could prove to be essential to someone who is new to operations or to an old hand who considers all subprocesses in his or her plant to be critical to production.

After matrixing and prioritizing each operational segment of your facility using this curve, it becomes apparent which of them are candidates for curtailment or total shutdown in case of a compromise to the utility being supplied to your plant.

For example, let's consider the potential for an electrical supply curtailment and compare two hypothetical but typical operational process segments: 1) a refrigeration compressor and 2) an instrument air compressor. These are similar enough to serve as examples of the use of this objective prioritization method. Each compressor is assigned a Process Criticality Factor and a Utility Demand Factor relative to the other subprocesses at your facility.

The assignment of a PCF to a subprocess can be as elaborate or as simple as you choose to make it. Generally, the more factors that are used to determine the PCF, the more useful the nomograph becomes as a prioritization tool; the priorities assume credibility. The ultimate credibility results from subjecting your major systems to true risk analysis. Only with this analytical tool can you be assured that the PCF you assign to a component has undergone truly objective scrutiny. However, rigorous risk analysis is a time-consuming task and many facilities are not staffed adequately to dedicate sufficient manpower to exploit its potential.

A less rigorous, less reliable, more subjective (hence, less defensible), but quicker method is to use the tables in this chapter as guides to assigning PCFs to your major pieces of equipment or to subprocesses that are discrete energy consumers:

***Table* 6–2 Guideline For Assigning PCF Points**

Need Provided by Subprocess	*Point Emphasis*
Safety	15–25
Quality	10–15
Productivity	7.5–10
Environmental Compliance	15–25
Hygiene	5–10
Morale	5–10
Economy	5–10
Other Factors	
Redundancy	(15–20)
Thermal Inertia	10–15
Ease of Restart	(15–20)
Likelihood of Property Damage	15–20

The assignment of a UDF is, on the other hand, a somewhat more straightforward task. It is commonly calculated in proportion to the utility demand a subprocess or piece of equipment has. The best way to assign UDFs is to create your own UDF/demand table based on the characteristic loads within your facility. For instance, if the largest discrete subprocess in your plant has an electric load of 2500 KW, you would want to assign it the maximum UDF (10) on your *electrical* UDF table. Then your ratio for assigning a UDF would be 0.004 UDF points per kilowatt. A typical UDF table for electricity may appear as follows:

Table 6–3 **Sample UDF Table For Electricity**

Discrete Subprocess	*Demand (KW)*	*UDF*
Bowl Crusher	2,500	10.0
Compressor, HVAC	1,500	6.0
Blender	1,300	5.2
Air Compressor, Instrumentation	500	2.0
Conveyor System	300	1.2
Agitator	200	0.8
Exhaust System Blower	100	0.4

Armed with this methodology, let's take another look at the two compressors we mentioned in the beginning. The HVAC compressor has been assigned an electrical UDF of 6 while the instrument air compressor has been assigned an electrical UDF of 2. The process criticality factor should have been calculated already as it would remain the same for no matter which utility interruption is being planned for. Considering the HVAC compressor first, the PCF numbers work out as follows:

Table 6–4 **Sample PCF For HVAC Compressor**

Need Provided by Subprocess	*Point Emphasis*
Safety	0
Quality	0
Productivity	7.5
Environmental Compliance	0
Hygiene	7.5
Morale	10.0
Economy	0

Table 6–4 Sample PCF For HVAC Compressor (continued)

Other Factors	
Redundancy	(0)
Thermal Inertia	0
Ease of Restart	(0)
Likelihood of Property Damage	0
PROCESS CRITICALITY FACTOR:	25

Now let's examine the instrument air compressor:

Table 6–5 Sample PCF For Air Compressor

Need Provided by Subprocess	*Point Emphasis*
Safety	11.5
Quality	15.0
Productivity	8.5
Environmental Compliance	0
Hygiene	0
Morale	0
Economy	5.0
Other Factors	
Redundancy	(0)
Thermal Inertia	0
Ease of Restart	(0)
Likelihood of Property Damage	10.0
PROCESS CRITICALITY FACTOR:	50

Plotting both of these factors on the nomograph for each compressor as shown in Figure 6–2, it becomes apparent that the refrigeration compressor is lower in priority to the plant than the instrument air compressor when considering curtailments of electricity. This is mainly due to the fact that the large compressor was assigned a low PCF because it was dedicated to comfort cooling only, not process cooling. In addition, it had a very high electrical demand compared to the instrument air compressor and, hence, was assigned a high UDF.

That was an example of the prioritization of an electrical load. In some facilities, steam load shedding is an important consideration. Steam is generated for process and space heating purposes. Some steam-generating plants produce superheated steam for electric power generation through steam turbogenerators. An increasing number of plants are installing combined-cycle gas turbine (CCGT) cogeneration systems complete with heat-recovery steam generators (HRSGs), or exhaust heat boilers. All of these produce steam. In most steam-producing plants, some of the steam produced is cycled back for internal use in the steam-producing cycle. In a Rankine cycle plant, a typical use for this internal steam is in the preheating of feedwater. Some or all of the feedwater pumps have direct-drive steam turbines. Many production machines or systems use direct-drive steam turbines; some enjoy electric motor backup.

Let's look at typical prioritizations involved for steam load shedding:

***Table 6–6* Typical UDF Prioritizations For Steam Load Shedding**

Discrete Subprocess	*Load (LBS/HR)*	*UDF*
No. 1 Turbogenerator	75,000	10.0
No. 4 Feedwater Pump	50,000	6.7
No. 1 Roll Machine	30,000	4.0
No. 2 Boiler ID Fan	25,000	3.3
Control Room Heating	20,000	2.7
West Vat Heater	14,000	1.9
East Emulsifiers	11,000	1.5
West Emulsifiers	10,000	1.3
North Thermocompressor	6,000	0.8
Admin. Space Heaters	4,500	0.6
Chlorine Vaporizers	3,000	0.4
Foaming Mixer	2,000	0.3

Need Provided by Control Room Heating	*Point Emphasis*
Safety	11.5
Quality	15.0
Productivity	8.5
Environmental Compliance	0
Hygiene	0
Morale	0
Economy	5.0

Table 6–6 Typical UDF Prioritizations For Steam Load Shedding (continued)

Other Factors	
Redundancy	(0)
Thermal Inertia	0
Ease of Restart	(0)
Likelihood of Property Damage	10.0
PROCESS CRITICALITY FACTOR:	50
Need Provided by Admin. Space Heaters	*Point Emphasis*
Safety	0
Quality	8.0
Productivity	8.5
Environmental Compliance	0
Hygiene	0
Morale	15.0
Economy	0
Other Factors	
Redundancy	(0)
Thermal Inertia	0
Ease of Restart	(0)
Likelihood of Property Damage	10.0
PROCESS CRITICALITY FACTOR:	41.5

Plotting both of these factors on the nomograph for each heating application as shown in the figure, it becomes apparent that the administration area heating is lower in priority to the plant than the control room heating. Again, we have objectively derived factors, the PCF and the UDF, indicating to us what is rather obvious: that process needs, in this case control room heating, outweigh comfort requirements most of the time.

Prioritization Objectivity

Although the relative importance of some processes to the overall facility mission may have been obvious to you at first glance, the prioritization of *many* operations may not be clearcut. That's when this rating method would prove very helpful by lending a degree of objectivity to your selection process.

As stated above, this technique is applicable to all kinds of process and energy streams. While you are performing risk analyses upon your energy supplies, this would be a good time to extend your analyses to include some utilities that do not supply energy to your facility, but which are energy-related and do consume energy. Some examples of these are water, steam, chilled water, and compressed air. You would probably be making a grave mistake not to follow through with this prioritization for all major incoming utilities subject to interruption.

Because prioritizing shutdowns of major components of your facility impacts the potential economics of your operations, you should be as objective and rigorous in your assigning a PCF as you can possibly be.

ECP OUTLINE & CONTENTS

The actual complexion of the plan will vary from plant to plant. But essentially, the plan is divided up into sections, parallel to the operating posture levels (I–IV) mentioned earlier. Each section specifies 1) Phase 1 (Identification), 2) Phase 2 (Evaluation), and 3) Phase 3 (Implementation)—needed *to move the energy posture of the facility to the next higher security level,* for example, Level I (Production Impeded—Code Red) to Level II (Production Erosion Imminent—Code Orange).

An outline for this plan follows:

***Table 6–7* Energy Contingency Plan**

Level IV—Normal Operations

No contingency actions required.

Level III—Production Curtailment Advisory

Phase 1 (Identification)

- Identify *specific* shortfall
- fuel inventory less than one week's supply
- energy-critical equipment with less than 10% capacity margin

Phase 2 (Evaluation)

- Review plan for widespread communication of shortfall
- Identify specific personnel involved
- Review "get-well" action checklist
- Review status-tracking procedure

Phase 3 (Implementation)

- Initiate widespread communication of shortfall advisory
- Notify specific personnel involved and assure receipt of notification
- Activate "get-well" checklist
- Implement status-tracking procedure
- Communicate status to specific personnel involved

Table 6–7 Energy Contingency Plan (continued)

Level II—Production Erosion Imminent

Phase 1 (Identification)

- Identify *specific* shortfall
- what production curtailment is imminent
- which fuel has less than two days' inventory
- what energy-critical equipment has no capacity margin

Phase 2 (Evaluation)

- Review plan for widespread communication of shortfall
- Identify specific personnel involved
- Compare cold-start costs to keep-warm energy rates
- Review "get-well" action checklist, including shutdowns and fuel-switching
- Assemble SOP shutdown priority list of major loads
- Review status-tracking procedure

Phase 3 (Implementation)

- Initiate widespread of shortfall threat
- Notify specific personnel involved and assure receipt of notification
- Advise specific personnel involved of cold-start/keep-warm recommendation
- Activate "get-well" checklist
- Implement status-tracking procedure
- Communicate status to specific personnel involved
- Submit SOP shutdown priority list to specific personnel involved

Level I—Production Impeded

Phase 1 (Identification)

- Maintain identity of *specific* shortfall(s)
- what production is curtailed
- which fuel is without inventory
- what energy-critical equipment is overloaded

Phase 2 (Evaluation)

- Review plan for widespread communication of losses sustained
- Identify specific personnel involved
- Compare cold-start costs to keep-warm energy rates
- Review "get-well" action checklist, including shutdowns and fuel-switching
- Assure production has been shutdown according to SOP priorities
- Review status-tracking procedure
- Determine on-hand inventory of possible fuel substitutes
- Determine current potential for fuel/power-switching

***Table 6–7* Energy Contingency Plan (continued)**

Phase 3	(Implementation) • Initiate widespread communication of losses sustained • Notify specific personnel involved and assure receipt of notification • Discontinue keep-warm if "get-well" time appears distant • Activate "get-well" checklist • Correct out-of-sequence curtailment • Implement fuel or power switching if possible • Implement status-tracking procedure • Communicate status to specific personnel involved

The medium-sized crumb rubber plant we began examining at the beginning of this chapter, which finally pieced together an energy contingency plan, faced an even more formidable danger than the energy supply disruptions and curtailments it provided for. That danger was complacency due to a "completed" plan. The plant manager was inclined to consider more immediate needs in his plant than the remote possibility of a now-distant contingency. Ultimately, the repeated advisories by the energy manager prevented the "plan" from being shelved. Instead, it became an "active" document, continuously undergoing changes to match the ever-volatile world energy situation and the process improvements constantly being made within the plant.

Working Tool

This ECP is not etched in stone; it is a working paper requiring review, updating, practicing, etc. It is worthy of mention that having a preplanned ECP saves implementation time, wards off panic, and tends to dampen panic-driven action which could spell future difficulty or disaster. The ECP should also contain action alternatives in case of unexpected circumstances which seem to crop up at the worst times.

A big reason for studying history, no matter which nation is examined, is to avoid repeating mistakes. But the trouble is, it doesn't seem to work very well. Instead, history repeats itself; it's absolutely cyclic. No aspect of life is immune, either; energy is certainly vulnerable to cyclical repetition. That includes the entire arena of energy supply contingencies. Recently, the world has been subject to repeated curtailments of oil supplies. The U.S. has experienced regional shortages of natural gas and heating oil at least once each winter for many years. Lessons learned? Yes! Shortages of energy have been subject to analyses at great length. Learning acted upon? No! Hence the cycles of energy contingencies continue.

The real danger in any plan, including energy contingency plans, lies in its *perceived completion.* Once a plan has a beginning, a middle, an end, an index, perhaps a title page, and a cover, it is perceived as being "complete" and the purpose for which it was compiled is perceived as having been satisfied. Our desire to insulate ourselves from a problem drives us relentlessly until we find a solution. It is only natural for us to seek a permanent solution so that the problem will be dealt with once and for all. A solution that requires revisiting very periodically is not an inviting solution at all. But it is precisely the solution required by such a fluid dynamic as the world energy supply situation.

7

Gaming the Contingency Plan

PRELUDE

Tom Danejack, a sanctuary ornithologist for the Coos Bay Wildlife Refuge on the coast of Oregon, pulled his new Jeep Cherokee into the parking lot of his local drug store late one spring Saturday afternoon. Tomorrow was Easter and he hadn't even bought the first marshmallow bunny or chocolate egg for his two little girls to celebrate with. He had no idea that the events of that day would not only prevent him from spending Easter with his family, but from seeing them for three weeks after that. That's not to say he didn't love his family; he was totally devoted to his wife Naomi and their two girls. In fact, it was his dedication to them that would drive him to the very limits of his physical endurance that spring of 1989.

He loaded up the Easter candies into the Jeep and drove to his office by the bay. The light on the answering machine was blinking. The office was open for a half day on Saturdays. Tom knew something was up, and Easter with the family might not be in the cards.

The call had been from Dr. Todd Rhyner at the marine university. A tanker full of crude oil had run aground somewhere in Alaska and was leaking pretty badly. Dr. Rhyner had gotten a call from the local Coast Guard commandant. There was a chance that he may have to go up there. They needed to talk right away.

Tom met Todd at the university and from there went to the U.S. Coast Guard office and were briefed in a little more detail, but not much. What they learned was that a supertanker had become disabled in a rare late winter storm and run aground on a reef in shallow water near a little place called Yakutat, home to about 200 people whose livelihood depended mostly on trapping and commercial fishing.

The area settlement was nestled in a rather pristine wilderness. Wildlife was abundant; water was crystalline. This was home to whales, porpoises, seals, sea lions, and sea otters. In the spring, millions of migrating sea birds, waterfowl, and shorebirds would flock to nest and feed on the ocean's buffet. Brown bear, deer, and bald eagles lived along the shores of adjacent Glacier Bay National Park. Aquatic life had to be abundant to support such higher orders of life, and it was. Hoards of Pacific salmon spawned upstream in brooks that emptied into the sea. There were also schools of herring, shrimp, and bottom-feeding fish.

That evening at the Coos Bay office of the Coast Guard, bits of information came in from other Coast Guard installations near the spill. Early on, it was apparent to the staff of the Alaska Division of Emergency Services that this was no little spill and that the ecological balance that characterized the area would be changed overnight, forever. After they'd asked for Coast Guard help, the Division of Emergency Services began canvassing the coast for expertise, especially from academia, to aid in mitigating some of the adverse environmental impacts of the spill.

So Dr. Rhyner and Mr. Danejack stood by for awhile. Finally, after a couple of hours of no further developments, they went home for supper and to pack in case one of the emergency response agencies answered their offer of help. It was Easter Sunday morning when they reported to the commandant's office. Since the time Tom had met with Todd the day before, an additional 50,000 bbl (2.1 million gal) of crude oil had leaked out.

That morning, the two men put together what they knew. Apparently, the spill had caught every major player involved by surprise. This included the EPA, FEMA, the Alaska Division of Emergency Services, the Alaska National Guard, the U.S. Coast Guard, and the owner of the vessel. From where they were, Todd and Tom could hear the confusion over the radio and the telephone conversations. No one seemed to be in charge.

In any contingency, response time is absolutely pivotal. It can prevent a major accident from becoming an unmanageable monster. While within the tanker, the 300,000 bbl (12,600,000 gal) of oil that would eventually leak out into the bay were contained by the walls of the hold. Once out in open water, however, it spread insidiously. This spill was never fully contained throughout the entire five months of initial response. It ended up becoming 25 miles wide and extended over 250 miles!

The two Coos Bay biologists played the waiting game for three days before they received a response to their offer of help. The call came from the oil company. They needed help salvaging scores of birds being coated with crude as they alighted on the black mire after miles of migrating back from southern regions.

When the men got there, they immediately set upon their seemingly insurmountable task. Thousands of birds were everywhere, black, shining, preening furiously and uselessly. They set up a washing station and began to work. The oil company assigned a couple dozen workers to help with the tasks. Other animals were brought in besides birds, such as otters and a few seals. The job seemed to have no end; the stench of death and oil permeated their nostrils incessantly. Their sleeping quarters in the nearby town of Yakutat consisted of a tent and a group shower. Still, the men, along with hundreds of others, worked feverishly for long hours.

If you work long enough in a single place, you can pick up information, even without asking. It seems that this entire event had been secretly feared by all the folks at Yakutat. When oil had been discovered in Prudhoe Bay, the people expressed their concern that their paradise's days were numbered. The State of Alaska, in 1977, before granting permits for the construction of the Trans-Alaska Pipeline, had been assured by written plan that the material, equipment, expertise, and manpower was in place and on continuous standby to respond to a spill.

The pipeline got built and the oil shipments began. Over the years, the track record of the system was admirable. *The lack of emergency situations slowly began to erode the sense of necessity for emergency planning.* The contingency support infrastructure began to dissolve. Any plans that may have been useful grew old and remained untested. Stockpiles of dispersant disappeared over time. The Emergency Response Team for pipeline contingencies was disbanded and the equipment dedicated to it either rusted or was cannibalized. And the U.S. Coast Guard hardly ever gave Yakutat a passing thought because it was not on any tanker route. But it was now on the minds of many Coast Guard personnel as the eyes of a concerned nation focused on the shores of Glacier Bay National Park.

With each passing day, more people showed up. Eventually, a command post was set up on the shore with a communication shed. The tanker had the remaining on-board crude pumped out while underwater welders set about making the hull floatable enough for towing. Gradually miles of oil boom material were dispatched, but the slick had spread too widely to hope for containment. So, boom material was placed as shields along the few remaining yards of unblemished beaches. The oil company located an oil skimming vessel in California which it put to work as soon as it came in.

It was four days before Todd met a representative of a government agency. She was the director of the Alaska Division of Emergency Services. It wasn't until the day after that representatives from all response parties could be seen together, meeting, brainstorming, and planning. Besides the press, there were folks from the EPA, FEMA, U.S. Park Service, Alaska National Guard, U.S. Coast Guard, Alaska Division of Emergency Services, Alaska Center for Marine Conservation, Alaska Department of Environmental Conservation, and some petroleum engineers from the oil company. It was as if a three-ring circus, a livestock auction, and a kangaroo court had been thrown together while someone took the last cup of coffee!

As a result, precious little progress was made. No one knew who should authorize a contract. So cleanup crews were still a paltry collection of desperate volunteers, mostly from the village. And they didn't have any cleanup equipment to speak of. Of them all, only the boom manufacturer didn't wait; the boom deployment crew said that their company's policy was to act first to keep costs low by keeping the problem manageable; the paperwork can always be sorted out later.

Eventually, things began to get done. The oil company had pledged to hire as many people as necessary to clean up the shorelines now blackened with a slick film. Of course, had swift action been taken in the beginning, a majority of the spilled oil could have been isolated from the shore. The governor of Alaska had turned over 500 National Guardsmen to work with the Coast Guard on accomplishing the logistics of the operation. They had heavy-lift helicopters for hauling the tons of trash and oil-laden material that accumulated daily.

All of this began to happen without the benefit of a plan. When the accident first happened, it didn't take much discernment, even on the part of the press, to realize that the plan that was on file (the one assembled in 1977) was useless. None of the names were current, nor were the phone numbers. Most of the equipment companies, consultants, and chemical services companies had gone out of business or moved. The state governmental staffing had changed drastically, and none of the organizational charts were accurate.

So most of the initial confusion that haunted this response in the spring of 1989 was a result of having to hammer out an entire contingency plan, complete with interagency coordination, within the time span of a few days. They had finally slapped something together while the tanker listed on its side. At least the oil company had arranged to begin pumping out the remaining crude from the stricken vessel to cut off the supply of contaminant.

This began as a minor energy incident and turned into a major environmental catastrophe. Before the end of the summer, it would be destined to blossom into a world-scale energy event.

As the cleanup operations continued, environmental solid waste began piling up at an exponential rate because of the expansiveness of the slick. Tons of oil-soaked wipes, disposable oversuits, tar globs, old buckets, broken hoses, hundreds of yards of oil-soaked boom material, and plain old garbage had accumulated despite the National Guard helicopters. Poor visibility on some days grounded the choppers. Meanwhile, scores of dead animals, mostly fish, otters, and birds, were bagged, tagged, and piled up. The oil company arranged to hire a fleet of barges to put the stuff in.

All four barges were filled and prepared for tow by an ocean-going tugboat. As it headed out for sea with the barges full of oily wastes, it passed downwind of the stricken tanker and the pumping vessel. Without warning, a strong wind came out of the west and blasted the shoreline. Oil booms were dislodged from their anchors and thick mats of coagulated oil emulsion and tar balls were pushed against the beaches of Glacier Bay National Park.

Then, a series of chain reactions began that no one had envisioned. The wind blew so hard that it dislodged enough of the tanker's hull to allow it to swing around to line up with the wind. In doing so, the tanker hull rammed into the hull of the pumper ship that was removing the oil. The sparks that ensued caught both vessels on fire. All personnel were immediately evacuated. Flaming pieces of material fell into the ocean, igniting the oil slick that surrounded the vessels. The fire spread along the top of the slick and reached the tug and four barges that were loaded with oily trash. The tug caught on fire and its captain gave orders to release the barges to drift in an attempt to save the tug and its crew. The barges drifted with the wind toward the shore.

As they drifted, one of the barges struck a reef and began to take on water. Meanwhile, the slick continued to burn and thick clouds of acrid black smoke filled the once-pristine air. No one noticed the barge sinking. It didn't take long for the barge to go completely under. Since it was lashed to the other three, it dragged them down, one by one. The hatches were secured but the sinking barge merely submerged the vents enough to allow water to begin filling the other barges, each in turn. And each in turn, they sank. All of the oil that had been absorbed by booms and wipes began to float once again toward the surface.

This nightmarish chain of events resulted in an intense polluting of the entire shoreline of Glacier Bay National Forest. But it didn't stop there. The burning slick was blown into the village of Yakutat where it actually caught the buildings next to the dock on fire. All pollution control efforts were halted as people scrambled in retreat from the burning slick, which had ended up in their lap. A little farther south, the burning slick set some spruce trees afire where they grew near the shore. The relentless west wind whipped the trees into a furious blaze. And so began one of the only recorded forest fires in the history of the park.

The confusion was spectacular. To his credit, the governor of Alaska insisted that the Coast Guard remain in charge of the maritime catastrophes. The U.S. Park Service was well-prepared to handle the forest fire. This was a well-rehearsed drill for them. Their execution of their firefighting mission was a welcome contrast to the mass confusion surrounding this ever-changing incident.

The governor of Alaska had had all he wanted with sloppy performance. With every hour that passed, he irretrievably lost some more of the finest features his state had to offer. Alaska, as he and the rest of Alaskans used to know it, was being eaten away. He immediately suspended operation of the Trans-Alaska pipeline. He publicly resolved that the suspension would last until he had in his hands, not only a decisive cleanup plan for the mess around Yakutat, but an up-to-date contingency plan from all entities involved in contingency operations in the state of Alaska, including (and this was crucial) signed commitments that drills will be conducted at least annually involving all of these entities.

The impact upon oil prices in the U.S. was unexpectedly rapid and severe. A full 25% of domestic crude oil production had been cut off almost overnight and the futures market panicked. The price for West Texas Intermediate went from $18.33 per bbl to $22.79. And it inched higher with each day's trading. The plans and commitments that the governor wanted could not be hammered out in a few weeks. This was a huge undertaking and the market knew that the curtailment would stretch on for many months, probably into the winter. But the governor felt justified. This calamity had mushroomed into a monster that was ruining, by the hour, the very wilderness that made Alaska the unspoiled frontier it was.

Of course, as the price of oil went, so did the price of all fuels, eventually. Natural gas sold for $3.45 per MCF within a week; No. 6 inched its way to $20.15 per bbl.; even coal enjoyed some resiliency in pricing, up to $53.15 per ton of Kentucky sub-bituminous by the time fall rolled around. At the pumps, motorists were greeted by $1.38 per gal gasoline.

Before the end of the summer, over 11,000 people were involved in cleaning the mess up; there were 1,029 vessels, 72 aircraft, 96 miles of boom material, 38 oil-skimming ships. The oil company spent over $2.5 billion, including funding the forest fire-fighting effort. And what did all this humanity, money, and equipment have to show for five months of effort by the time fall came around? Of the over 800 mi that had been polluted, 200 remained to be cleaned; 17,000 birds had died, along with 700 sea otters, 80 deer, 10 porpoises, four killer whales, and two seals. Tom Danejack and Todd Rhyner worked long hours through the mosquito-filled, oil-stenched summer, tediously washing birds and other animals off and preparing them for relocation if they survived. They were ready to resume normal life back down in Oregon by the time the cleanup efforts were suspended for the winter.

The unfortunate reality was that most of the millions, maybe billions of dollars could have been saved. The impact of this entire series of incidents could have been cut to less that a tenth of what it ended up being. How? Simply, *response time.* If the initial response time had been minimized, the oil could have been contained and not allowed to gush uncontrollably for so long while so many people wondered what to do. The ideal response to this whole disaster should have consisted of a quick assessment followed by a rapid dispatch and mobilization of mission-specific crews, material, and equipment. No hesitation; no reasoning out of anything; no arguing over who's in charge. In short, *the whole response should have been executed in proper sequence, with minimum timing, and with the precise amount of material and equipment.*

But it's not enough to have a good plan.

Planning Is Not Enough

Having a good contingency plan is like providing assembly line workers only with instructions as to how to assemble a car. It's not enough, not if you want them to do it right and in the minimum amount of time. They need training and practice, practice, practice. In the real world, practice is better known as "experience." It's not something that comes from books; it comes from doing something "hands-on."

The timely and proper execution of any contingency plan, such as the Energy Contingency Plan (ECP) explained in Chapter 6, is directly proportional to the time spent practicing with it. Have you ever watched a college marching band or a military precision drill team go through their routine? How do they become as good at something so complex, involving a multitude of separate individuals? Practice; drill, drill, drill.

But isn't a drill involving so many outside contractors, agencies, and companies rather expensive? Ask the CEO of the oil company whose vessel ran aground how expensive a drill of a good plan would have been.

Also, how can someone tell how really "good" a plan is? Your ECP may look great sitting there on the shelf with its brand new date on the cover. But is it worth the paper it's printed on? How can you tell? Only by putting it through its paces. Subject the plan to some imagined catastrophes and see if it provides the framework so necessary to respond to the calamity. See if you can still maintain production at your facility by using your new ECP.

Gaming

The consideration of "what if" scenarios, either by staging the circumstances, or by merely discussing possibilities, is an essential ingredient of an effective energy contingency plan.

This artificial staging of an event is called **gaming** in the contingency profession and should be performed in increasing frequency relative to the degree of probability that a catastrophic event will occur. All critical endeavors should be gamed. It is merely an efficient analytical tool used to realistically test the contingency plan under fire.

Gaming provides the means to compare your current state of readiness with future energy supply contingencies. It is an analytical tool that can illustrate how uncertainty and human decision making under stress can affect the unfolding of events. At its core is conflict among decision makers that have competing goals. This is good. Real life is like that, and in an energy contingency situation this conflict intensifies.

By putting your energy contingency plan through the rigors of a true-to-life simulation exercise, one excellent result is a platter full of insights into the strengths and weaknesses of your ECP. Among the other benefits to implementing your plan under the aegis of a simulation is that it familiarizes all involved parties with procedures that they would use in an actual emergency. It allows the players to *network* with each other under crisis conditions. It builds rapport and comaraderie among actual people who would be dealing with each other in a real emergency. Gaming tests communication networks and helps to identify administration, data, and information requirements. It forces the participants to appreciate the time and space requirements of major emergency tasks. It can also reveal how adroit you are at handling the media.

Some of the products of successful gaming are an improved Energy Contingency Plan, an increased state of readiness among participants and leadership, a set of well-established routines applicable in multiple scenarios complete with checklists, guidelines, and standard operating procedures.

The primary advantages to gaming are to illustrate the ramifications of uncertainty and to demonstrate the criticality of human decision making under pressure. Although this decision making is hampered by pressure, gaming the ECP gives individuals much needed practice to take the panic edge off this process. The gaming process 1) forces the participants to view different aspects of problems, 2) allows participants to entertain a multitude of assumptions, 3) allows conflict to occur between decision makers with competing goals, 4) allows for testing of the validity of assumptions, and 5) serves as an effective way to communicate your concerns for energy contingencies to your employees, the public, and the government.

The *most valuable product* of repetitive gaming, for all of the people that truly participate, is the *unfolding of a process* of how to react to a real energy emergency, a process including a communications net, recall procedures, information systems of all sorts, etc. These are all things that can be applied when any crisis arises for which there is no set plan.

Repetitive gaming tends to fine-tune the ECP, exposing shortcomings, honing it down to an effective and working document. It is impossible to learn true energy emergency management in a totally calm, nonemergency atmosphere. Study of the ECP should take place under calm conditions, but *practice should be performed under realistic conditions*. This is what most large agencies concerned with emergency planning do. Agencies such as the DOE, NATO, DOD, IEA, etc. game their plans.

Levels of Intensity

Gaming takes place in two degrees, low-level and high-level. Low-level gaming is seminar-type training with roundtable discussions of various scenarios. But leafing through a three-ring binder is no substitute for hands-on participation in a high-level simulation where the people involved in energy emergency response learn to cope with the stress of crisis and make decisions under those conditions. The experience of being bombarded with truckloads of data and other information while being forced into intense decision making is something you can't get with low-level gaming.

High-level gaming involves a multitude of participants from many agencies who would be involved in an energy emergency. This latter degree of gaming represents a sizable investment because it involves large amounts of manpower and equipment for hours on end.

Because of the money involved, it is probably a good idea to engage the services of a gaming consultant to work out the appropriate scenarios and objectively evaluate the effectiveness of your personnel and their use of the ECP.

To maximize the effects of participation in high-level gaming, it is necessary for the participants to suspend cognizant disbelief, much like what it takes to get the full enjoyment from a novel. A novel is not read for factual information, but to study and appreciate human character. Similarly, the high-level gaming of an energy supply contingency is used to get at the root of your ability to respond adequately. It is not so important that a crisis similar to the one being gamed could really happen; what's important is how the participants behave under pressure; what networks were absent; what routines could make the response more timely (recall the oil spill in our story at the beginning); what standard operations procedures were conspicuously absent; etc.

Not a Performance Evaluation

Gaming should be considered a learning experience for you and your employees, not some sort of performance evaluation. It is imperative that upper level decision makers at the facility be involved in this learning experience. Employees who participate must begin to feel comfortable with the emergency procedures in the ECP; they should begin to understand the analytical processes involved in choosing options; they should become familiar with the time-based crisis dynamics of an energy contingency and the strain put upon normal decision-making abilities; they should, by their participation, display support for energy contingency planning.

Out-of-Date ECP

An out-of-date plan could prove worse than no plan at all because it lends a false sense of security to a situation and may even result in a waste of time pursuing ineffective avenues of contingency response. This happens time and again. It's exactly what happened in the Alaska spill. An annual gaming of your ECP should reveal shortcomings.

Contingency Management Consultant

If you don't know where to start, you should contact a professional contingency management consultant. This consultant should not only be retained to annually review your plan, but should be contracted to design a full-scale gaming operation of the ECP. The cost of his or her fees would not even scratch the surface of the costs of a botched response to an energy emergency.

One final case for gaming your plan is that annual updates or gamings make ideal opportunities for the involvement of new participants, such as new employees, in the process.

Interdepartmental Participation

The involvement of representatives from all of the organizations tasked for a response in your ECP should not be difficult to obtain because most of those entities know that gaming needs to be an integral part of their job; it's just not done often enough. They would probably surprise you with their willingness to participate in either level of gaming your consultant recommends. After all, this unique type of training allows them to fumble, patch up frayed information networks, make the wrong decisions, and witness the adverse impact of those bad decisions, without having to actually pay the price of a wrong decision during a real crisis.

8

Energy & the Environment: A Reluctant Synergy

PRELUDE

One of the world's most gruesome technological environmental spectacles was given birth in mid-January 1991. It grew furiously, waiting for no bit nor bridle. The horror of its wrath was the subject of worldwide conversation and media coverage for months. How to not just stop it, but how to kill it formed the focus of many tactical planning sessions by experts from many countries.

Before the Iraqi army had fled Kuwait, they executed an asset-denial tactic on a scale heretofore unseen and for tactical purposes heretofore unknown. The statistics are staggering. More than 500 oil wells were ignited, spewing flames 200 feet in the air and billowing smoke two miles high. It was said that their roar sounded like hundreds of jet aircraft taking off. At the height of the disaster, *five million barrels of oil were wasted each day,* converted into nothing less hideous than a half million tons of aerial pollutants. Black rain fell in Saudi Arabia and Iran while black snow fell in Kashmir 1,500 miles to the east. To add to the senseless waste and ruin, the Iraqis blasted oil storage facilities and pipelines while emptying a half dozen loaded oil tankers (especially hauled in for this). They managed to spill *six million barrels of crude oil* into the Persian Gulf.

The resultant slick covered 600 square miles and ruined 300 miles of coastline.

It is ironic that these tactics could have been so thorough. The activities of terrorist Iraq were not a series of clandestine operations directed against an unsuspecting foe. This occurred at the tail end of a war; this part of the world had been the worldwide focus of attention for many months. But happen it did, and caught the pollution-fighting capabilities of that area unprepared for a catastrophe of that magnitude. The countries to be involved in the cleanup operations had also been involved in the war. This served to hinder the environmental response from governments intent on military victory.

The wholesale decimation of the fragile ecology encompassing the geography of the Middle East was only one of two edges of a desperate sword wielded by a people hungry for a last shred of pride as they ran from the enemies they had chosen. For with this onslaught of incomplete combustion, furiously blazing for no purpose whatsoever, the Iraqis denied billions of joules of fossil energy assets to a world fighting each day with batts of insulation and fluorescent lamps to scrimp and save a watt here and there, unyielding in their conviction that an unburned BTU of fossil fuel helps to keep the environment healthy for humankind. The other edge of the sword, dripping with entropy, was that the Iraqis' reduced these efforts at energy conservation to insignificance with little effort on their part; fossil fuel got burned, anyway.

Inextricable Linkage: Environmental Responsibility, Energy Management, and Economics

Environmental responsibility and energy management are inextricably linked. These two foci of technology are often seen at odds, but that is only a delusion. As you look for options to keep your operations powered, you'll find less and less license to choose without taking into consideration a host of environmental considerations. You could argue ad infinitum about how right or wrong this is. That's beyond the scope of this book. Your very customers, the consumers of your products, have defined a market for economical products of high quality with minimal environmental impact.

Also, you'll find another "E-word" lurking in the equation. That word is "economics." Because the simple fact is that if you can maintain quality production safely while using less energy, you've enhanced the profitability of your company while lessening the impact on the environment. The wise use of energy, true energy management, is one of the most cost-effective efforts you can take to increase the energy security of your facility because of this three-way relationship.

Second Law of Thermodynamics

The linkage between energy and the environment is not surprising, considering that the Second Law of Thermodynamics holds true from Greenland to Argentina and from Puerto Rico to Baja California. The entire ecosystem we live in is moving from a highly available energy state to one of lower availability, due to pervasive entropy. The only reversal to this decline comes from our sun. This seems to indicate that your long-term strategy for energy security should be one that is generally based on using fossil fuels in the near term while you gradually move your energy base to one powered by renewable energy sources.

Consider the case of imported oil. Recall that the U.S. is so dependent upon oil imports that a moderate disruption in supply wreaks widespread pandemonium upon the way of life we have carved out for ourselves. Thus far, the security of an adequate supply of energy for the world's enterprises has withstood the test of time. A cycle involving synthetic fuels, foreign vs. domestic production debates, etc. has developed since WWII. At each supply disruption, we see these debates. And at each debate, the deciding factor against nonforeign oil is the economy of foreign oil. Substitutes for foreign oil, be they domestic oil, synthetic, or alternate energy forms, are always quashed by the economy and availability of foreign oil.

The New Oil Nemesis

Now, a new nemesis to foreign oil has arisen. Frankly, this paradigm has begun to change the way all fossil-based energy is viewed by the people of today. It gained significant strength in the aftermath of Desert Storm. It's not called "supply," or "lack of reserves," etc. It will make a true case for nonforeign energy. It could boost the production of the synthetic fuels industry, and certainly the renewable energy industry. Its major strengths are its tenacity, its endurance, and its pervasiveness into all aspects of our society. It is here to stay; it is not something that will go away, or is "out there" waiting to be dealt with on other soils. This has involved a host of people in the past to get us to where we are, and now involves every man, woman, and child in this age and those to come. Many people in our society perceive that we are choking in our own waste, particularly by our use of certain types of energy. The perception is that if the environment has no future, neither does mankind. If mankind has no future, neither does the need for energy.

Your first inclination as an industrial manufacturer may be to fight this perception of people. Many industrial giants have tried. Look at their success rate. For example, when someone mentions the name "Exxon," what do most people think? And that's *after* they spent $1.2 billion in Alaska! It is probably not good business to confront environment situations without a heaping helping of mollification. Why is this so? Why have rational people embraced this attitude so ideologically? Some psychologists have concluded that concern for the environment has supplanted the innate drive within people to worship a higher being than themselves. In other words, God's name, to some people, is Environment.

The reasons why concern for the environment has become so pervasive is beyond the scope of your trying to assure that your facility doesn't starve for energy. However, it is essential to recognize the depth of conviction characteristic of these people ... they are your customers; they keep your doors open.

World Direction

If you are like most people, you share in the belief that the world is progressing toward a more valuable state as a result of the steady advance of technology and the accumulation of knowledge. We seem so cognizant of all around us that we marvel at the shortsightedness of our ancestors. Our lives seem to be lasting longer and we seem to be suffering less from physical ailments. The world seems to be headed in the right direction.

Actually, that way of thinking had its roots in the thoughts of men who lived some 300 years ago. These men, among them Isaac Newton, Francis Bacon, Rene Descartes, John Locke, and Adam Smith, fashioned an outlook upon the world that differed from the ancient philosophies of the Greeks. As a whole, the ancient Greeks held that the happenstance of the world lacked order, and nature was unpredictable, as were the notions of mankind. The ushering in of the mechanical age by men such as the above held that the world was indeed quantifiable, predictable, and behaved in an orderly fashion. Buried within the beliefs of the new mechanical age philosophers was the notion that people largely improved the value of their lives by the accumulation of wealth, and technology was the perfect tool to allow this advancement. Of late, this view is coming under more scrutiny and enjoying less popularity because of the Second Law of Thermodynamics. This warrants some explanation.

Laws of Thermodynamics

The study of thermodynamics contains some observations of the relationships between matter and energy which are called "laws." The First Law of Thermodynamics can be summed up thusly: energy can neither be created nor destroyed. Now, energy can be changed, and frequently is changed, from one form to another. But that is not a violation of any law of thermodynamics. The Second Law of Thermodynamics is a bit more revealing. Simply stated: energy flows from a more useful state to a less useful state, from a state of order to an equilibrium state or a state of chaos. This character of disorder has a name; it's called "entropy." Entropy is a measure of the amount of energy which is no longer useful for performing work.

Pollution and Entropy

Every time something happens on the Earth, some amount of energy has become less available for doing work. Ecological pollution is a rather obvious highly entropic state with material eternally blended. You've probably observed that heat always flows from a hotter body to a colder body until the hot one cools off some and the colder body heats up some. Eventually, both will attain the same temperature and the flow of energy stops. This equilibrium state is a state in which internal energy exists, but because there is no temperature difference, no work can be done.

This observation or "law," that energy moves from a high "available" state to a lower "unavailable" state, actually permeates everything in our world. For instance, when hydrogen gas is burned or oxidized, water and heat results. The mass of oxygen and the mass of hydrogen used in the combustion process is still there; they have not been destroyed. In fact, the hydrogen and oxygen in the water can be separated, but it takes some energy, some heat. What about the heat that was given off when the two chemicals were burned? That would be perfect—except it has long since been dissipated into the overall energy of the room. So, it will take some additional energy (some *available* energy) to return the two chemicals in water back to their original elemental state. There they'll stand ready to recombine and release the heat that was used to separate them when they were water. But you see, each time we burn these two chemicals, we've got to go to another energy source to separate them. The reason for this is couched in the Second Law of Thermodynamics, otherwise known as the Law of Availability or the Law of Entropy.

Since entropy is the measure of work now unable to be used for doing work, when energy flows cease because equilibrium of pressure or temperature has been reached, or when heat is dissipated in the air, it is said that entropy has *increased.*

And it's inescapable. It's what makes perpetual motion impossible. Many people are of the impression that almost everything can be recycled and reused over again as many times as we choose. That is a mistaken notion. Recycling requires the expenditure of available energy from somewhere for collecting, sorting, transporting, and processing of the recycled material. This use of available energy increases the entropy of our overall environment. The more we pull from this pool of available energy, the more energy is dissipated, the more entropy is increased, the less capable our environment is to perform work because it is closer to its final state of equilibrium. Isn't that spooky! But look around you. Have you ever seen a violation of the Second Law of Thermodynamics? If you think you have, don't approach the U.S. Patent Office; they refuse to patent perpetual motion devices.

Sources of Available Energy

But there's a piece of this equation that hasn't been mentioned before. It's kind of central to this chapter. It has to do with having a pool of some available energy for doing work. On this planet, aside from our launching some satellites into deep space and a deposit of cosmic dust and meteorites, there is no exchange of matter with the rest of the universe. What we have is what we're stuck with matter-wise, like it or not. With respect to energy, though, things are a bit different. Here on Earth we have two sources of available energy, energy useful for performing work: 1) renewable and 2) nonrenewable. Both are limited in the amount available for exploitation.

Renewable energy derives its entropic decrease from the influx of solar energy. It can be a direct capture of the sun's rays, or a harvest of stored solar energy in such forms as biomass, hydro-energy, wind, or the thermal gradients in the oceans. The renewable energy in stored form is limited in the amount we can capture for exploitation, so much so that it can, with overexploitation, become similar to nonrenewable resources in availability. The amount of solar energy we can directly capture is almost limitless, but harvesting is severely restricted due to the varying rates and patterns with which it arrives on earth. So, it too has limitations.

Nonrenewable sources of energy are technically renewable, but only over a geological timeframe which, for humans with relatively short lives, may as well be classified as infinity, which makes these resources nonrenewable, for all practical purposes. Ever since the Industrial Revolution, when we initially turned to coal, we have become more and more dependent upon nonrenewable forms of energy. Nonrenewable-based resources, such as plastic, have practically supplanted renewable resources, especially wood, as basic construction materials.

Presently, our economy is fueled mostly by nonrenewable energy. Consequently, the transformation of this energy from one form to another, from one level to another, is only resulting in making more nonrenewable energy unavailable for work. There is no scarcity of fossil fuels. There are huge amounts of coal buried in the United States and in Russia. The supply of natural gas is more dependent on the transmission capacity of the pipelines rather than proved reserves in the ground. Not many days can go by without your hearing about oil and the economy of the Middle East. There still remains much oil under the eastern sands.

Well, if there's not much of a supply problem with nonrenewable fuel sources, why should we be worried about increasing the entropy of our world? What is the reason we should be concerned that with every transformation of a drop of oil into heat, water, and carbon dioxide, we increase the overall entropy of the world we live in? As long as you have assured that your facility receives an adequate supply of energy so that it can produce its product, you have done your job, right? In fact, there seems to be enough fossil fuels left to power the endeavors of many generations after you. So, you should even be able to sleep better at night, knowing that you haven't denied the generations to come the energy that is their birthright, correct?

Is the only reason that you are concerned about wasting energy the fact that it is uneconomical? Do you ever get flashes of guilt when you are forced by the nature of your manufacturing process to waste energy, even when it is more economical than shutting down production?

The Concept of Sinks

As strange as it may seem, all of these questions can be answered within the context of the Second Law of Thermodynamics. Basic thermodynamics reveals that in order for energy to produce work, it must flow from a state of higher availability to one of lesser availability. This lower energy state must be accessible to the higher energy state in order for the energy flow to be initiated. In other words, a process must have a "sink." A thermal process must have a heat sink or a level of *thermal* equilibrium; a chemical process must have either a heat sink or a level of *chemical* equilibrium, or both. As life goes on, and we take nonrenewable energy sources and move them through the thermal processes we need to maintain the standard of living we've created for ourselves, we are actually gradually closing the gap between the higher energy state found in fossil fuels and the equilibrium state found in the great sink we call the environment. So, there's the rub; there's the "ratchet clause" in our energy contract. We are, energy-wise, headed down a one-way street; and all technology can do for us is to get us to the end of the street faster than when we started.

What does this equilibrium state look like? In the movie *Beverly Hills Cop*, the character played by actor Eddie Murphy disabled an unmarked police car with a banana by merely stuffing it into the tailpipe. In the movie *Star Trek VI*, the character Uhura states, while looking for the "Achilles heel" of an invisible enemy spaceship, "... the thing's gotta have a tailpipe." Virtually every industrial complex is replete with stacks that are necessary to keep the processes from choking to a standstill in their own waste. All dynamism requires a flow of energy. All work requires a "tailpipe" to give the process access to its equilibrium sink. When the sink is full, the process has to stop. The final equilibrium state resembles a full kitchen sink (pun intended). And the environment which we are using as the process sink for all of our "tailpipes" is filling up with each btu of fossil fuel we burn.

So, what's the solution? What's the point? Not to use energy that is nonrenewable? This is a rather unreasonable suggestion in light of the fact that the cultures of most of the world are now embrace or are in the process of embracing lifestyles that depend almost exclusively on fossil fuels. But the fact is that the "sink" is filling up; our environment is becoming increasingly saturated with insidious pollutants, making it more difficult to deposit any more. And the people of the world are noticing this. Even our laws are gradually sliding "bananas" into the smokestacks of our industries in an attempt to slow this slide into the highly entropic equilibrium state within which no work is possible.

This may all sound rather philosophical to you. It should; no serious study of the Second Law of Thermodynamics has ever neglected the philosophical ramifications of how all of this highly available energy got put into the universe in the first place. Philosophy doesn't keep a plant running (or does it?). What are the ramifications for your facility in light of this entropy consideration? Does the entropy law hold at your plant or are you immune to its effect?

THE FILLING SINKS

Almost daily you can read in the newspaper about a huge company laying off hundreds of people, closing plants, divesting assets, etc. If you were an "economic doctor," what would your diagnosis be of what is wrong? Remember in the good old "boom" days when environmental accountability was not even a passing thought? Corporations were growing like teenagers, full of energy, anxious to try their hand at anything, full of "vip 'n' vinegar." They could do no wrong. The government was supportive, their employees were supportive, their neighbors were supportive. They grew; they gained momentum; they grew some more.

But most importantly, they acquired a mindset. And just like people, once they developed a mindset, it became harder to change as they grew older. Despite the fact that sometimes a change of attitude is called for as times change, large mega-corporations had a hard time changing the policies that had worked so well in the past to build them up to the huge agglomerations that they are today. Part of that mindset included the opinion, almost a creed, that *just about any action that resulted in progress was a worthwhile action.* Little attention was given to the effect on ethnic populations, who or what was downstream, or what organism was lost to the world. But what appears as short-sightedness today wasn't the fault of the corporations; corporations were only responding to the demands of their customers. And when the corporations were growing up, customers paid little attention to long-term effects of present-day actions that resulted in present-day prosperity. In other words, corporations had assumed the mindset of their customers.

SPONSORS OF CHANGE

The attitudes of customers are changing. Corporations are having a hard time following that change. The "voice of the customer" is growing louder in favor of environmental accountability. Today's customer is a rather cognizant individual who is becoming aware that the true cost of a modern product is not necessarily reflected in its price. As customers make their voices heard, corporations are trying to respond. And respond they must, to survive. Because corporations in and of themselves are shells; they are doomed without support from their stockholders, their employees, and their customers. The behemoths have a harder time about it than the smaller companies because they lack agility. Quite frankly, some are so big that they are paring down by laying off employees and selling "nonstrategic" business units to regain that agility they once enjoyed while they were growing in response to their customers' wants. There is definitely some churning going on in the business world.

This impacts you at the plant level. If your facility is part of one of these large corporations that is rather loathe to change, you are impacted more directly. If that is the case, constructive change, based upon the voice of your customers, may be incumbent upon your plant so that it can work its way upstream to your central corporate workings. If your facility is part of a smaller corporate structure, change may be coming to you by way of policy rewrites in response to the voice of the customer.

What happened to the support that corporations enjoyed when they were growing up? What happened to the voice of the customer, the employee, and the neighbor of big business? Instead of unqualified support, they all now seem to be demanding different things: the customer wants more value and quality in products and services; the employee wants more self-improvement training and quality family time; the neighbor wants aesthetics and a cleaner environment in which to rear a family. Although each of these types of people whom the corporation depends upon seem to be demanding different things, they are really demanding different aspects of the same thing—accountability. They are demanding corporate accountability in exchange for measured support; it's as simple as that. The days of corporate license are no longer supportable by the people because the materials, energy, and the environment won't, in turn, support that style of growth at any cost; the people have seen the cost/benefit ratio and have become disappointed in it.

What's actually happening is a result of the people feeling the effects of the Second Law of Thermodynamics. The "sink" is filling and they're having difficulty treading water, polluted groundwater, that is, or ultraviolet rays from a lack of ozone, or greenhouse gases from too much combustion and not enough rain forest, or garbage, or smog, or acid rain, or strip mining, or clear cutting, or species destruction, or whatever combination they perceive is a result of man's insatiable appetite for limited resources.

Nature's Closed Loops

People are learning that the processes of Nature are actually closed loops. Take a trip out west and visit a grove of giant sequoias. Most of those trees are 175 decades old while the life of the average corporation floats between three and four decades. Each tree is a system of chemical processes much like your facility. Why have they lasted so long? How can you make your facility more like a giant sequoia so that it will last more than 30 or 40 years? Each sequoia is operating in a closed loop, mirroring ecological systems world-wide. By finding this closed loop balance, each ancient tree has attained sustainability.

Most technological processes invented by man do not operate in a closed loop; there is no balance with the core character of life on Earth and, consequently, no basis for sustainability for man's processes. Recall that the proponents for the new mechanical age in which we live advocated the accumulation of wealth as the avocation of mankind by the use of technology. They postulated that the Earth and its creatures are the property of humankind, a gift from the supernatural, if you will. The embrace of this paradigm had man standing outside of nature with nature as a playing field which demanded domination, control, and manipulation. As technology has grown, so has the assumption been reinforced that there should be essentially no limit to humanity's power over nature. These man-made processes, results of the Industrial Revolution, became linear. On one end of each of these linear processes is the exploitation of seemingly inexhaustible resources; on the other end of each of these linear processes is the "sink" for the depositing of waste. But because the resources are exhaustible and because entropy always increases (and the sinks get full), there can be no sustainability to these processes such as the sequoias enjoy. The result of the wholesale adoption of linear processing is that unsustainable development has generated wealth and comfort for only about one fifth of the world's population.

Paradigm-Shifting

The "nitty-gritty" for your plant is energy security to maintain production. As you should be gaining awareness by now, as long as you fight to keep the processes in your facility linear, you have sealed their fate into unsustainability. Unsustainability is diametrically opposed to maintainability. If you wish to begin taking steps to lock in your facility's energy security, you will need to begin taking steps to first, reduce its unsustainability, and second, increase its sustainability. That begins with a major paradigm shift on your part. You must begin discarding the philosophy of the ushers of the Industrial Revolution: the accumulation of wealth by the use of technology is not the avocation of mankind, nor is it the purpose of technology; the Earth and its creatures are not the property of humankind; man does not stand outside of nature; nature is not a playing field which demands domination, control, and manipulation; and there is a limit to humanity's power over nature.

Once you've begun to make these attitudinal adjustments, you'll need to begin looking for those energy and material resources that will close the loop around your currently linear process. This means that you will need to begin weaning your facility from nonrenewable fuels and raw products and begin exploring which *renewable* resources you could incorporate. This should further enhance your paradigm shift toward nontraditional means of meeting the desires of your customers. What that means is that you should begin climbing out of the rut of thought that limits your ability to meet the desires of your customers. Your current rut limits your response to the voice of the customer to only those products you currently manufacture. Listen to your customers. Become bold enough to think of possible ways to meet the needs of your customers that don't require the huge capital investments and ravenous appetites for materials and energy that your current products do.

What all of this does is make quite a case for 1) energy conservation, 2) renewable energy use, and 3) protective guardianship of those systems which make renewable energy available to us. Of the foregoing three, the one which is immediately available to you at the least cost is energy conservation. And this energy resource is doubly cost-effective in that it inches your facility closer toward sustainability and saves you money on top of that.

The favorite method we've developed to deal with potentially serious catastrophes is the purchase of insurance. People think it very rational to pay premiums such that in case disaster strikes they or their survivors will be better off than if they'd not purchased insurance. And no one considers it strange for insurance purchasers not to complain if disaster fails to strike. The pursuit of sustainability can be viewed as purchasing an insurance policy. Doggedly ferreting out opportunities for energy conservation and hunting for opportunities to use renewable energy and materials to feed your operation will further the closing of your linear processes into loops. The closing of your processes into loops allows you to share in the sustainability enjoyed by the sequoias. And what if the Earth's environmental sink never gets full and greenhouse gases don't build up, and ultraviolet rays don't attack us through an ozone hole, etc.? The price you will have paid for the sake of sustainability will have yielded secondary benefits or dividends such as increased efficiencies, improved human health, and more widely distributed prosperity.

What have you got to gain? What have you got to lose?

9 Summary

World Energy Dynamics

The energy situation for countries who are net importers of energy is extremely complex and is in a constant state of flux. But at the core of this situation is a precarious dependence being forged upon increasingly undependable sources of this energy. The manager of an industrial facility needs to provide for an ample supply of energy to his plant, and that is getting to be harder to assure as times goes on.

An energy supply is a core necessity in the viability of your plant; it flows throughout; it's almost everywhere you look; it's indispensable; it's much like blood is to the human body. Fortunately, the body has bone marrow which manufactures the necessary supply of blood cells. The modern industrial facility is unlikely to enjoy such a captive supply of energy. So the fuel buyer casts the plant's energy needs into a pool with other energy users, all of whom scramble for a constantly dwindling supply of domestic oil. They then carry their efforts abroad to see what they can get internationally. This is a rather simplistic view of what a plant manager faces trying to power his facility. What's the solution?

There is a case for industrial plant survivability from an energy supply standpoint. The ability to run your plant depends less upon your corporate affiliation as it does upon your locality and its proximity to different energy sources. Energy security is very regional. Energy security is centered around the integrity of your supply routes and methods of energy conversion.

Increasing the energy supply security at your plant is an involved process which encompasses many steps and facets. It is an effort that needs to penetrate the rank and file from top to bottom to be successful. It is an effort which is serious enough to demand unwavering support by upper management and broad enough to involve your people closest to the process. It is a preventive program, not a cure-all to be applied after the fact. It is a detailed, sometimes tedious, dogged effort that needs to be ram-rodded by people with the right "stuff." It is an ongoing effort, since nothing abroad nor locally is static.

Some authorities may advocate a different series of steps toward energy contingency planning than is offered in this book. There is no best universally applicable sequence of steps. But it is better to start somewhere rather than not at all.

Prioritization

One of the first things you will need to do to start this program rolling is to prioritize your in-house energy-consuming operations. You focus upon the primary mission of your facility, which may be the production of one or more products, and you examine how directly each individual energy-consuming operation is linked to your primary production. You look at how directly an operation within your plant affects the safety of your people, the quality of your product, the sheer amounts of energy it consumes and the rates at which it consumes it. You look at its importance to the process stream, the ease at which it can accept a substitute energy source, and the time it would take to repower it with that other source.

One of the best tools to help assign a priority to these processes is risk analysis. It can lead to those processes which are essential to the operation of your facility. These critical operations are then assigned a "Process Criticality Factor" which is directly proportional to the criticality of the operation of your facility.

Physical Security

The physical security aspect of energy contingency planning is a broadly ranked subject in itself. Although this book treats a narrow scope of security, i.e., energy supply, making your energy supply physically secure goes hand-in-hand with making your plant physically secure. The majority of security measures to be considered will center on protection of facilities against humanly perpetrated actions. But the precise degree of protection depends on what type of person is taking physical action against your plant and where this person will strike. You need to know your enemy and your plant before you plan your defense.

Some of the best advice is to engage the services of a security consultant. These people can analyze your current security posture relative to your deficiencies and the current technology that can rectify those deficiencies. Your major contribution to all this is to assign a value to certain processes and energy supply streams so that you can get an idea of how much of an investment in these new security measures you want to make.

Some industries are directed by the government as to how much security to use, such as nuclear power plants, ammunition depots, and some military installations. But most facility managers enjoy a healthy amount of latitude as to how much security to employ.

Some semblance of security already exists at most facilities. This is mainly to protect companies from theft, vandalism, loss of propriety, and to insure the safety of their employees. What usually happens to plant security is a gradual degradation to a level which allows the guard to stay in his or her shack and wave at people entering and leaving the plant. What aggravates this situation is that nominal gate security does not come close to protecting your facility from industrial terrorism or sabotage.

It would be impractical to make an industrial plant an impenetrable fortress. The amount of security you provide your plant will depend on many things. One of the very first considerations you should have is, "Who is my enemy?" You need to determine the type of individual(s) who are likely to breach your perimeter.

Your security methods will depend upon what energy stream you are protecting, its accessibility, the type of person likely to attack it, and how the attack will likely occur. You will probably conclude that a wide variety of potential saboteurs and targets exists. Even if you were to beef up your effort and launch an all-out bid for physical security for the energy choke points at your plant, the enthusiasm of your personnel may soon falter. You may even incur a large expense and be perceived as overreacting.

After you have attempted to identify the types of potential enemies you are facing, you compile a list of vulnerabilities or "choke points." These are the locations, major pieces of equipment, or process bottlenecks in your plant. These will be the focal areas upon which to concentrate your security efforts.

The types of measures you should already have in place include such things as perimeter fencing topped with three strands of barbed wire, area lighting, and a main gate guard shack manned around the clock. Depending upon the likelihood of aggression, you may want to have more than this.

Energy Management

After you've prioritized major aspects of your operation and examined possible improvements to your physical plant security, another way to decrease your vulnerability is to simply depend on less energy. This is cost-effective and independent of the type of energy your plant requires. Reducing energy consumption per unit of product is a vital step and should be done concurrently with the other initial energy isolation measures mentioned before. You need to get and keep your plant as energy-lean as possible.

You can't control what you can't measure. So the management of your energy consumption requires measuring all of the energy streams throughout the plant, such as electricity, natural gas, waste fuels, coal, petroleum coke, lignite, and fuel oil. You will also need to measure major water flows into, within, and out of the facility, such as raw water, process water, wastewater, and cooling water. These water flows are excellent indicators of anomalous energy usage. Besides measuring current flow rates, you will need to gather historical data of these flows for truly in-depth analysis.

You get a handle on all this with what is referred to in the industry as an industrial energy audit. It consists of two degrees of investigation. The first is sometimes called a "walk-through" audit. The second is an in-depth energy analysis which usually results in an energy balance survey of your plant. To get to this point requires a methodical, relentless, and tedious effort into what resembles detective work: looking for clues from available data to emerge with strategic, long-range energy management opportunities (EMOs), usually more capital-intensive than what the tactical or walk-through audit reveals. The EMOs from the in-depth analysis should more than likely be long-range in scope and would become part of your strategic energy management plan.

The auditing and compiling of the plan can be performed by your in-house technical staff if you have one, or a qualified consulting firm. Just bear in mind that this planning will become part of your energy security planning, which could be a factor in the survivability and profitability of your plant. As such, these people, especially if they're in-house staff, need your support, your input, and your involvement because of the many interplant departments it will affect.

The results the audit yields should not be viewed as stagnant facts. Too many things are going on which impact the economics and security justification of an EMO: process technology, energy and equipment prices, personnel, internal plant/process modification, etc.

Flexible Power Options

One of the most effective methods, albeit capital-intensive, of securing a power supply for your plant is to give it an option for using more than a single fuel. Although it is possible to enjoy a wide variety of these options, capital considerations usually allow for duplication of a particular piece of equipment if anything at all. But duplication appears to be adequate and is certainly superior to having a single source of process energy.

Besides hardening your energy supply against disruption, multifueling allows you to take advantage of fluctuating fuel prices among the different avenues into your plant. It is a very effective energy security measure, but because it is very capital-intensive, you will need to use it for other reasons to enhance its cost-effectiveness. Otherwise, you will have capital tied up in idle facilities. One thing to remember with multiple energy sources is that the facilities and equipment which allow you this option need to be maintained and periodically checked for operability.

Besides the advantages listed above, you have the laws of chance in your favor; it is highly unlikely that more than one major energy source will be disrupted or become overpriced at the same time. Also, multifueling, except in automatic switching of electrical sources, does not encompass a high degree of technical complexity.

Waste Streams

Waste streams take on significance in plant operations, especially in securing your energy supply, because one way to enhance energy security is to use less energy in your processes. Waste streams from a plant contain different qualities of recoverable energy. The more energy you can recover from waste streams, the less primary energy you must buy. Waste streams also serve as indicators of your process efficiency. You can use these streams as such. If you can decrease the temperature or flow of a waste stream from your facility without sacrificing product quality, you have enhanced the energy security of your plant.

The Contingency Plan

Because of the dynamics of industrial operations, the facility modifications, process changes, financial variances, and personnel turnovers, it is essential to the maintenance of energy security operations at the plant to nail down these operations into some sort of logical, implementable plan. This planning by necessity is divided into the short-range, tactical planning and the long-range, strategic planning. Piecing the plan together should start with the long-range view, the strategic plan. In this way, by the time your planning works its way into the details of tactical planning, these details will be part of an overall direction for your plant and will likely avoid conflict with other tactical planning directions.

One of the primary considerations when beginning this overview of where you are and where you'd like to be is, How does your plant fit into the contingency planning of your major energy suppliers? Your facility is on a "hit list" that your major energy suppliers will use to target your plant with cutoff at some point in crisis. You need to find out what that point is for each energy supply stream. Press your supplies to reveal what conditions would force this curtailment. This will make your planning more meaningful.

The strategic portion of your plan should include a general description of the changes of operations necessary to isolate your plant from a major supply interruption. The strategic portion of your energy contingency plan should describe the major capital improvements necessary to do this. It should also call out the needs for fuel redundancies needed and intended.

The tactical portion of your plan should include a more definitive posturing of particular assets and detailed operational parameter changes which are to support the overall direction of your planning. Redundancies of equipment necessary for these changes and any multifueling intended should be specified if it can be done within a short time frame.

Gaming the Energy Contingency Plan

The consideration of "what if" scenarios, either by staging the circumstances, or by merely discussing possibilities, is an essential ingredient of an effective energy contingency plan. This artificial staging of an event is called "gaming" in the contingency profession and should be performed in increasing frequency relative to the degree of probability that a catastrophic event will occur. All critical endeavors should be gamed. It is merely an efficient analytical tool used to realistically test the contingency plan under fire.

The primary advantages to gaming are to illustrate the ramifications of uncertainty and to demonstrate the criticality of human decision making under pressure. Although this decision making is hampered by pressure, gaming the ECP gives individuals much needed practice to take the panic edge off of this process. The gaming process 1) forces the participants to view different aspects of problems, 2) allows participants to entertain a multitude of assumptions, 3) allows conflict to occur between decision makers with competing goals, 4) allows for testing of the validity of assumptions, and 5) serves as an effective way to communicate your concerns for energy contingencies to your employees, the public, and the government.

Repetitive gaming tends to fine-tune the ECP, exposing shortcomings, honing it down to an effective and working document. It is impossible to learn true energy emergency management in a totally calm, nonemergency atmosphere. Study of the ECP should take place under calm conditions, but practice should be performed under realistic conditions. This is what most large agencies concerned with emergency planning do. Agencies such as the DOE, NATO, DOD, IEA, etc. game their plans.

Gaming takes place in two degrees, low-level and high-level. Low-level gaming is seminar-type training with roundtable discussions of various scenarios. But leafing through a three-ring binder is no substitute for hands-on participation in a high-level simulation where the people involved in energy emergency response learn to cope with the stress of crisis and make decisions under those conditions. The experience of being bombarded with truckloads of data and other information while being forced into intense decision making is something you can't get with low-level gaming.

High-level gaming involves a multitude of participants from many agencies who would be involved in an energy emergency. This latter degree of gaming represents a sizable investment because it involves large amounts of manpower and equipment for hours on end. Because of the money involved, it is probably a good idea to engage the services of a gaming consultant to work out the appropriate scenarios and objectively evaluate the effectiveness of your personnel and their use of the ECP.

Gaming should be considered a learning experience for you and your employees, not some sort of performance evaluation. It is imperative that upper level decision makers at the facility be involved in this learning experience. Employees who participate must begin to feel comfortable with the emergency procedures in the ECP; they should begin to understand the analytical processes involved in choosing options; they should become familiar with the time-based crisis dynamics of an energy contingency and the strain put upon normal decision-making abilities; they should, by their participation, display support for energy contingency planning.

An out-of-date plan could prove worse than no plan at all because it lends a false sense of security to a situation and may even result in a waste of time pursuing ineffective avenues of contingency response. An annual gaming of your ECP should reveal shortcomings. A professional contingency management consultant should not only be retained to annually review your plan, but should be contracted to design a full-scale gaming operation of the ECP. The cost of his or her fees would not even scratch the surface of the costs of a botched response to an energy emergency.

One final case for gaming your plan is that annual updates or gamings make ideal opportunities for the involvement of new participants, such as new employees, in the process.

Interagency Participation

Part of your energy emergency planning needs to involve your local and regional energy officials, law enforcement agencies, and other disaster response officials. A flow of communication with them during your planning process not only results in a more effective plan, but it opens and maintains interagency personnel familiarity which is conducive to the effective prosecution of an ECP. The regional energy emergency plan will more than likely impact your plant. You need to get a copy of this plan and incorporate it into your own. The Federal Emergency Management Agency can help with this.

Legislation

As an entity responsible for a large portion of the local and regional economy, your government representatives should be depended upon to consider your views and needs regarding energy supplies for your facility. You should formulate opinions and business policies about a multitude of energy security topics and let your legislators know of these positions. Typical issues are energy pricing, environmental issues, state energy emergency plans, national energy policy, utility purchases of nonutility power, electrical grid access, and demand-side management programs.

Conclusion

The current world energy situation makes the subject of energy security a timely topic. It is not a simple subject, as we have seen; it is multifaceted. It embraces the movement throughout the plant to "squeeze" the most useful production out of each kilojoule or kilowatt-hour you buy ... to get "energy-lean." It requires inventory through the auditing of your energy use. It calls for prioritizing your operations.

It includes protecting your facilities from sabotage, vandalism, and terrorism or catastrophes of a natural origin.

Novel energy supply strategies are considered for options such as multifueling and waste stream energy exploitation. All of these exploratory studies result in strategic and tactical realms to the planning of your response to energy emergencies. The result is your plan, in writing, but always dynamic—never static.

The key to your successful energy contingency planning is involvement of all your people across the board and of respective government agencies. Involve them in auditing, recommending, strategy sessions, and gaming the plan. Invite them to participate with your corporation in the influence of their legislators.

So much depends upon continued operations at your facility. So much of your continued operations depends on your supply of energy. Don't let happenstance rule that energy supply; there is a lot you can do to reduce the risk of interruption and assure that your plant is in it for the long haul, despite the volatility of energy.

Bibliography

Ahearn, Joseph A. "The FAST START Plan." *The Military Engineer,* January/February 1989.

Badolato, Edward V., Julius Bleiweis, Julie D. Craig, Horace W. Fleming, Jr. *Hurricane Hugo: Lessons Learned in Energy Emergency Preparedness.* Clemson University: The Strom Thurmond Institute, 1990.

Badolato, Edward V. *Federal-State-Industry Roles in Energy Emergency Preparedness.* Clemson University: The Strom Thurmond Institute/U.S. Department of Energy, Summer 1988.

Badolato, Edward V. *Gaming an Energy Crisis: Energy Preparedness and DOE's Emergency Exercises.* Washington, DC: U.S. Dept. of Energy, March 1988.

Badolato, Edward V. *The Vulnerability of the U.S. Oil and Gas Transportation System.* Williamsburg, VA: Energy Safety Council, 1990.

Bracher, B.L. *Strategies for Reducing Natural Gas, Electric and Oil Costs—Chapt. 94, Utility Risk Analysis.* Atlanta: Fairmont Press, 1989.

Boraiko, Allen A. "Earthquake in Mexico," *National Geographic,* May 1986, pp. 655–675.

Bruckner, Donald G. "As the Walls Come Tumbling Down," *The Military Engineer,* March–April 1990, pp. 43–46.

Bunker, Robert M. "Military Engineers and Disaster Relief: Hurricane Hugo," *The Military Engineer,* March–April 1990, pp. 36–39.

Canby, Thomas Y. "After the Storm," *National Geographic,* August 1991.

Carpency, Frank M., and R.C. Rittenhouse. "Power Plant Security." *Power Engineering,* September 1980.

Coltharp, David R. "Designing Buildings Against Terrorists," *The Military Engineer*, August 1987.

Denton, Hon. Jeremiah A., Jr. "Domestic Terrorism—The Nature of the Threat." *State and Local Government Responses to Domestic Terrorism*, Conference Proceedings, June 1986.

Directorate of Readiness. *AF Pamphlet 93-XX, Volume I, Pre-Disaster Planning.* Washington, DC: U.S. Air Force.

Eguchi, Ron, and Ian Austin. "The Lessons of the Loma Prieta Earthquake," *Elements (Dames & Moore)*, Vol. 1, No. 19, 1990, pp. 14–15.

Farrell, LTC William R. *Naval War College Review*, "Terrorism: Why, What, and When," January/February 1986.

Findley, Rowe. "Eruption of Mount St. Helens," *National Geographic*, January 1981, pp. 1–74.

Fleming, Horace W., Jr., et. al. *Regional Differences, Common Concerns.* Clemson University: The Strom Thurmond Institute, 1988.

Fletcher, Robert P., Jr. "To the (Search and) Rescue," *The Military Engineer*, March-April 1992, pp. 4–8.

Frydenlund, Marvin G. "Understanding Lighting Protection," *Plant Engineering*, December 13, 1990, pp. 54–58.

Funk, Ben. "Hurricane!" *National Geographic*, September 1980, pp. 346–353.

Fuqua, C.S. "Computerizing Catastrophe," *Business Alabama Monthly*, September, 1991, pp. 33–36.

Gibbons, John H., Peter D. Blair, and Holly L. Gwin. "Strategies for Energy Use," *Scientific American*, September 1989, pp. 136–143.

Guidelines for Hazard Evaluation Procedures, 2nd Ed. New York: American Institute of Chemical Engineers, 1992.

Hall, Stephen K. "Oil Spills at Sea," *Pollution Engineering*, December 1989.

Haller, Mike. "Oil Spills: Alaska and Delaware Assist in Cleanup," *National Guard*, November 1989.

Horwich, George, and David Leo Weimer (Eds.). *Responding to International Oil Crises.* Washington, DC: American Enterprise Institute, 1988.

Jeffery, David. "Yellowstone: The Great Fires of 1988," *National Geographic*, February 1992, pp. 255–273.

Johnson, Wray R. "Coping with Violence Abroad," *Air National Guard Civil Engineering Association Newsletter*. Washington, DC, March/April/May 1991.

Lee, Douglas B. "Tragedy in Alaska Waters," *National Geographic*, August 1989.

Mashburn, William H. *Managing Energy Resources in Times of Dynamic Change*. The Fairmont Press/Prentiss-Hall, 1989.

McDowell, Bart. "Eruption in Colombia," *National Geographic*, May 1986, pp. 641–653.

Miller, James Arnold. *Report of the Conference on International Energy Security*. Washington, DC: U.S. Dept. of Energy, 1988.

Miller, Peter. "Tornado!" *National Geographic*, June 1987, pp. 691–715.

Nichols, Alan B. "Emergency Response Keyed to Local Planning." *Journal WPCF*, March 1988.

O'Brien, James T. "Guarding Buildings with CCTV/Motion Detectors." *Consulting/Specifying Engineer*, June 1988.

Patterns of Global Terrorism: 1990, Washington, DC: U.S. Dept. of State, April 1991.

Peterson, Grant C. "FEMA and Hugo," *The Military Engineer*, March–April 1990, pp. 40–42.

Rifkin, Jeremy. *Entropy: Into the Greenhouse World*. New York: Bantam Books, 1989.

Ruckelshaus, William D. "Toward a Sustainable World," *Scientific American*, September 1989, pp. 166–172.

Stickney, Wallace E. "FEMA vs. Mother Nature," *The Military Engineer*. January–February, 1992, pp. 9–14.

Sununu, Hon. John H. "A Governor's Perspective: The Emerging U.S. Energy Crisis." *Report of the Conference on International Energy Security*, May 1988.

Tarics, Alexander G. "Earthquakes and Our National Defense," *The Military Engineer*. March–April 1990, pp. 31–35.

Terrorist Group Profiles. Washington, DC: U.S. Dept. of Defense, November 1988.

Trout, John E. and Curt P. Betts, *The Military Engineer*, July 1988, "Security Engineering."

Turner, Wayne C. (Ed.), *Energy Management Handbook*, 1982, John Wiley & Sons, Inc.

Vesely, W.E., F.F. Goldberg, N.H. Roberts, D.F. Haasi. *Fault Tree Handbook*. Washington, DC: U.S. Nuclear Regulatory Commission, 1981.

Ward, Fred. "Dominica," *National Geographic*. September 1980, pp. 357–367.

West, Bing. *Midwest Regional Energy Emergency Simulation*. Falls Church, VA: GAMA Corporation.

Yergin, Daniel. "How to Design a New 'Energy Strategy,'" *Newsweek*, February 11, 1991, p. 44.

Yergin, Daniel. *The Prize*. New York: Simon & Schuster, 1991.

Young, Steven S. "Terrorism and Environmental Security." *Pollution Engineering*, April 1988.

West, Bing, *Midwest Regional Energy Emergency Simulation*, 1988, GAMA Corporation, Falls Church, VA.

Index

I

K

L

M

O

P

Q

R

S

T